CHEMICAL MICROANALYSIS
USING ELECTRON BEAMS

CHEMICAL MICROANALYSIS
USING ELECTRON BEAMS

I. P. JONES

THE INSTITUTE OF MATERIALS
1992

Book No. 523

Published in 1992 by
The Institute of Materials
1 Carlton House Terrace
London SW1Y 5DB

© 1992 The Institute of Materials
All rights reserved.

British Library Cataloguing-in-Publication Data
Available on application

ISBN 0 901716 06 5

Printed and made in Great Britain by
The Bourne Press Limited, Bournemouth

Preface

When I was commissioned to write this book, I suddenly became an avid - if rather surreptitious - reader of all the previous essays in the genre. Studying these earlier efforts, it became very clear to me that no two authors had the same opinion concerning what a 'microanalyst' needed to know. There are the instrumental buffs, the computer freaks, the applications enthusiasts... and so on. Of course, I too have had to decide for whom this book is written, and with what in mind. It is part of a series commissioned by the Institute of Metals[*] as first introductions to various aspects of quantitative analysis. I have attempted to develop a description of electron beam microanalysis based on simple physical principles. This book, therefore, does not contain detailed information on instrumentation or on alternative quantification procedures. The emphasis throughout is on understanding what is happening, and why.

I have consciously repeated some important explanations and descriptions in slightly different ways at various points throughout the text. I believe this is an important, although often underrated, aspect of teaching. For those who open this book to check quickly on a particular formula or set of data, I offer my apologies for any irritation caused as a result. In fact the most used and useful part of any book is its appendices and thereby generally is it judged! I have tried to include all important data in Appendix A and formulae in Appendix B.

The example calculations are all designed for pocket calculator. Obviously for regular data analysis a microcomputer would be more appropriate and to reproduce, for example, some of the computed graphs in the text is a useful exercise. Why bother with calculation by hand when most modern microanalytic equipment is supplied with dedicated computer and proprietary software? I believe passionately that using such software without an understanding of what it does and how it works not only invites but guarantees disaster. Nine times out of ten the initial results are wrong for one reason or another. Usually they are outrageously so and it is important to be able to recognise this and understand why. Hence this book.

By the time I finished writing I was left with an absolute loathing for units, the product of many hair-tearing weekends looking for some small but crucial inconsistency between different versions of the same equation as published by different authors. One upshot of this is that I have

---

[*] Of blessed memory: now the Institute of Materials.

scrupulously used Système International (SI) throughout. This has resulted in some fairly unfamiliar looking units - for example $m^2\ kg^{-1}$ for mass absorption coefficients instead of $cm^2\ g^{-1}$ - but I think the gain in consistency easily outweighs this.

Acknowledgments

At Birmingham I am part of a large team of people working on and with various aspects of chemical microanalysis. I cannot overemphasize how important this has been in the preparation of this book. In particular, Malcolm Hall helped me very much with the practical and quantitative aspects of bulk specimen analysis. I wish also to thank Tony Burbery, without whose wide and detailed technical knowledge I would frequently have foundered, and whose equanimity has miraculously survived my periodic assaults on his microscopes.

I am grateful to the following for allowing themselves to be inflicted with parts or all of the earlier versions of this book, and for their helpful suggestions, to the more polite of which I generally acceded: Ray Bishop, Tony Burbery, Graham Cliff, Peter Goodhew, Alfonso Ngan and Alan Nicholls.

In any case, ... quorum aemulari exoptat neclegentiam potius quam istorum obscuram diligentiam. dehinc ut quiescant porro moneo et desinant male dicere, male facta ne noscant sua.[*]

I have often read, in prefaces to other books, fulsome acknowledgments to immediate family, with apologies for the enormous disruption caused, irritability of the author, etc. How could such a simple thing as writing a book be responsible for all this? Now I know, and make in my turn my due acknowledgments and apologies.

1989-1992
Birmingham and, somewhat more enjoyably, Falmouth and Florence

---

[*] Terence: prologue to the Andria. A rough translation is: 'People with ultrathin windows shouldn't throw high energy electrons'.

CONTENTS

Chapter 1   Introduction   1

Chapter 2   **The interactions of electrons and X-rays with solids**

2.1	Preamble	7
2.2	Overview	7
2.3	Single electron excitation	12
2.4	Plasmon excitation	28
2.5	X-ray and Auger electron emission	30
2.6	X-ray absorption	36
2.7	X-ray fluorescence	42
2.8	Inelastic collisions between beam electrons and nuclei : bremsstrahlung	45
2.9	Elastic scattering of the beam electrons by atoms	47
2.10	Elastic scattering of the beam electrons by crystals	54

2.11	Summary of Chapter 2	56

Chapter 3 **Microscopes and spectrometers**

3.1	Scanning electron microscope / electron probe microanalyser	57
3.2	Transmission electron microscope	62
3.3	Scanning transmission electron microscope	66
3.4	Energy dispersive X-ray spectrometer	66
3.5	Wavelength dispersive X-ray spectrometer	82

Chapter 4 **Working out the compositions**

4.1	Thin film X-ray analysis	92
4.1.1	Spectrum $\rightarrow$ X-ray intensities detected by Si (Li) crystal	95
4.1.2	X-ray intensities $\rightarrow$ chemical composition	102
4.1.2.1	Very thin specimens	102
4.1.2.2	Thicker specimens	107
4.1.2.3	What about *very* thick specimens?	121
4.2	Bulk specimen X-ray analysis	121

4.2.1		Extracting the X-ray intensities (including any attenuating effect of the detector)	123
	4.2.1.1	EDX detectors	123
	4.2.1.2	WDX detectors	124
4.2.2		X-ray intensities → chemical composition (the 'ZAF' corrections)	125
	4.2.2.1	The stopping power correction	128
	4.2.2.2	The backscattering correction	136
	4.2.2.3	The absorption correction	139
	4.2.2.4	Fluorescence by characteristic X-rays	146
	4.2.2.5	Fluorescence by continuum X-rays	154
	4.2.2.6	Summary of formulae used for ZAF corrections	160
	4.2.2.7	An example of a full calculation	161
	4.2.2.8	Final comments concerning chemical microanalysis of bulk specimens	170

Chapter 5   Some miscellaneous topics

5.1   Beam spreading   173

    5.1.1   Beam spreading in transmission specimens   173

5.1.2	Beam spreading in bulk specimens	175
5.2	Low energy X-rays	176
5.3	Alchemi	179
5.4	Electron energy loss spectroscopy (EELS)	180
5.5	Some valedictory comments	188

References 189

Appendices

Appendix A 195

A.1	Some useful constants and formulae	196
A.2	Atomic numbers and relative atomic masses	197
A.3	Characteristic X-ray energies	198
A.4	Fluorescence yields	203
A.5	Partition factors	206
A.6	Mass absorption coefficients and ionisation energies for X-rays	207
A.7	The logarithmic integral	217

Appendix B

    B.1   Symbols and Acronyms   218

    B.2   Summary of formulae   224

Appendix C   **A reminder on atomic structure**   233

Index   237

Chapter 1 Introduction

The chemical analysis of small volumes of material using electron beams forms an attractive subject for a book for two reasons, one intellectual and the other technological. Firstly, the nature of the various interactions between a beam of electrons and a solid, and the subsequent events, form a coherent and satisfying body of knowledge. Secondly, an understanding of the ways in which these interactions are exploited for chemical analysis gains from their being studied in conjunction with each other. In this introductory chapter I shall set out in broad terms what this book is about and, just as importantly, what it is *not* about.

The principle of electron beam chemical microanalysis is that a narrow beam of electrons is fired into a solid. The electrons interact with the solid. Their energies and directions change and the solid itself emits X-rays and electrons (Fig. 1.1). By studying the effect of the solid on the incident electrons, and the various emissions from the solid provoked by the electron

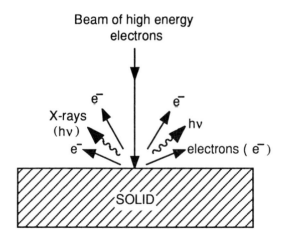

Fig. 1.1 A beam of high energy electrons strikes a solid specimen. X-rays and electrons are given off. Analysing the energies and intensities of the X-rays enables the chemical composition of the solid to be determined. Some of the electrons (the 'Auger' electrons) may also be analysed to give a surface chemical composition. This is dealt with in another book in this series (see footnote to next page).

2 Chemical Microanalysis Using Electron Beams

beam, we can deduce the chemical composition of the solid. The major part of this book is concerned with the X-rays emitted by the specimen and how we analyse them. Although some of the emitted electrons - the 'Auger' electrons - define the chemical composition of the specimen just as much as do the emitted X-rays, their generally low energies mean that they come from a layer very close to the surface of the specimen. 'Auger spectroscopy' is a surface analysis technique which will be described in another book in this series.*

As a minor theme, this book will deal with how the changes in energy of the beam electrons can be used for chemical analysis (Fig. 1.2). The changes in *direction* of the electrons will not much concern us. This is because the changes in energy give much more specific information about the chemical composition of a specimen than do the changes in direction. The latter form the basis of electron microscopy.

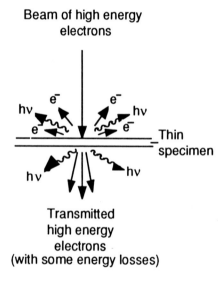

Fig. 1.2 If the specimen is thin enough to transmit the electron beam, the energy losses of the beam may be measured and used to deduce the composition of the specimen.

---

*'Quantitative surface analysis for materials science' by G.C. Smith.

Thus the essence of this book is as follows:

> A narrow beam of electrons is directed at a solid. The energies of the resulting X-rays are characteristic of the elements present. By measuring their intensities the chemical composition of the irradiated volume may be determined.

What instruments are used for this type of analysis? There is an important division to be made at the outset, between instruments designed primarily to exploit *transmission* of the incident electrons through the specimen and those designed for the examination of *bulk* specimens. The two types of instrument we shall be involved with, illustrated schematically in Fig. 1.3, are:

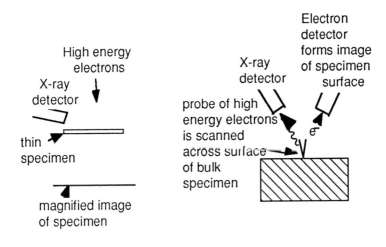

(a) Transmission electron microscope

(b) Scanning electron microscope / electron probe microanalyser

Fig. 1.3 The two basic instruments. In the transmission microscope (a) the electrons pass through a thin specimen and form the image all at once. In the microanalyser / scanning microscope (b) the image is produced sequentially from the surface of a bulk specimen.

4   *Chemical Microanalysis Using Electron Beams*

(a) The *transmission electron microscope*, where the thin specimen appears half-way down a column of electromagnetic lenses. A magnified image of the interior of the specimen is produced on a screen at the end of the microscope. The instrument is usually operated in *fixed beam* mode where the whole picture is formed at once, but may be operated in *scanning* mode like the second type of instrument described immediately below. It is possible, and may be useful, to examine the energy losses of the transmitted electrons.

(b) In the *scanning electron microscope** a small probe of electrons is scanned across the surface of a bulk specimen. The electrons reflected or excited from the specimen are detected and their intensity displayed synchronously on a cathode ray tube to form an image of the surface of the specimen. A very similar instrument is the *electron probe micro-analyser*. The difference between the two is purely one of emphasis. The 'microprobe' is designed primarily for microanalysis, whereas the scanning microscope is primarily for imaging.   I shall be dealing with both instruments in this book - largely interchangeably.**

When does one use a transmission, and when a bulk, specimen?

The advantages of transmission specimens are:

(a) Spatial resolution is better: the incident electrons are scattered about less in a thin transmission specimen than in a bulk specimen.

(b) The correction procedures which have to be applied to the measured X-ray intensities to derive from them the chemical composition are much simpler for thin specimens.

---

*A misnomer, because transmission microscopes can operate in scanning mode.
**The blurred and increasingly undetectable difference between electron microprobe analysers and scanning electron microscopes was nicely illustrated to the author recently when a technical engineer arrived to install a new scanning electron microscope. On discovering that a wavelength dispersive analyser (see later) was to be fitted he carefully unscrewed and detached the plate reading '...... scanning electron microscope' and replaced it with one reading '...... electron probe microanalyser'.

(c) It is possible to derive useful information from the energy losses of the transmitted electrons.

The advantages of bulk specimens are:

(a) Many more X-rays are given off by a bulk specimen than a thin specimen. The composition sensitivity of the technique is therefore much higher.

(b) Specimen preparation is simpler.

(c) The influence of surface films resulting from specimen preparation, whose chemical composition is different from the true composition, is far less for bulk specimens.

So far I have said nothing about how the X-rays are monitored. There are two fundamentally different methods: *energy dispersive* detection and *wavelength dispersive* detection (Fig. 1.4). In energy dispersive detectors the X-ray buries itself in a silicon crystal. Some of the energy it loses is expended in converting bound electrons to free electrons. The pulse of conductivity this gives rise to is measured and the energy of the X-ray deduced. This can be done thousands of times per second. In a wavelength dispersive detector X-rays of a particular wavelength are first isolated from all other wavelengths by Bragg reflection from a monochromating crystal and are then counted in a proportional counter. In energy dispersive analysis all energies or wavelengths are analysed for at the same time and the analysis can be built up very quickly. The much slower wavelength dispersive analysis has, however, better sensitivity, particularly at larger X-ray wavelengths (lower energies), and is the preferred technique under certain circumstances. The lower efficiency of the wavelength dispersive technique currently limits it to bulk specimen analysis, where there are far more X-rays given off. Energy dispersive analysis is used for both transmission and bulk specimens.

The energy losses of the electrons transmitted through a thin specimen may be monitored by a spectrometer placed at the end of a transmission electron microscope.

That completes the brief survey of electron beam microanalysis promised for this chapter. In Chapter 2 I shall discuss the basic physics of the

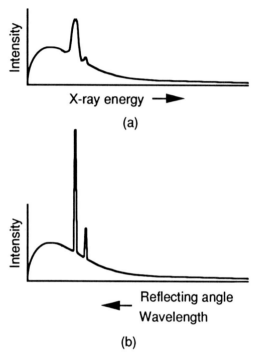

Fig. 1.4 Output from two types of X-ray detector: (a) energy dispersive and (b) wavelength dispersive. Notice that the x axis or abscissa is really the same in both cases: as the X-ray photon energy increases, the X-ray wavelength decreases. The sharp peaks show *which* elements are present; their heights show *how much* of the element is present.

interactions between the electron beam and the specimen and the subsequent interactions between the generated X-rays and the specimen as the X-rays pass through the specimen on their way to the detector. In Chapter 3 the experimental equipment is described: the electron beam instruments and the X-ray detectors they incorporate. Chapter 4 contains an account of the correction procedures necessary to convert X-ray intensities into chemical compositions. Finally Chapter 5 presents some miscellaneous topics, including a brief account of electron energy loss spectroscopy.

Chapter 2 The interactions of electrons and X-rays with solids

2.1 Preamble

The theme of this book is chemical microanalysis via the X-rays excited by a small beam of electrons. To be able to interpret the X-ray information it is necessary to understand the interactions between the electrons and the specimen, and between the excited X-rays and the specimen. This is the subject of Chapter 2.

2.2 Overview

When a fast electron enters a solid, there are four possible interactions, all electrostatic in origin, which are important in the present context. They are illustrated in Fig. 2.1 and are as follows:

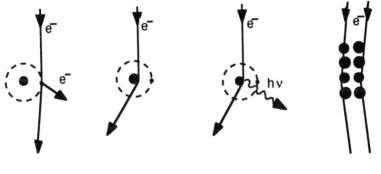

(a) Collision with electron
(b) Scattering by atom
(c) Scattering by nucleus with emission of photon (bremsstrahlung)
(d) Scattering by crystal

Fig. 2.1 The four interactions of a fast electron with a solid.

8 Chemical Microanalysis Using Electron Beams

I. *Inelastic collisions with electrons.* Here the interaction is with the bound, atomic electrons within the solid. It is usually referred to as 'inelastic' because internal energy is transferred to the atom from the kinetic energy of the incoming electron. There is an electrostatic repulsion between the negative charge of the incoming fast electron and the negative charges of the electrons resident in the solid. Some of the kinetic energy of the fast electron is transferred to the resident electrons. There are two possibilities (see Fig. 2.2): in a *single electron* collision, one electron is promoted from

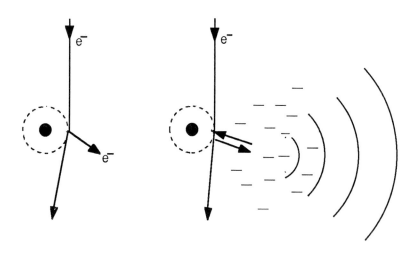

The atomic electron is ejected.

The atomic electron is restored to its orbit by the repulsion of the neighbouring electrons (the 'plasma'). These in their turn are repelled. The resulting wave is a 'plasmon'.

(a) Single electron   (b) Plasma

Fig. 2.2 The two types of electron-electron interaction: in (a) the interaction is with a single atomic electron; in (b) it is with many interacting solid state electrons.

its current energy level in the atom to a higher energy level, which is nearly always unbound - i.e. the electron is ejected from the atom. This leaves a space in the electronic structure of the atom, to which we will return presently. The second possibility is a *collective* interaction, where the excited electron is either free (a conduction electron), or towards the outside of the atom. Its displacement is restrained by neighbouring electrons of the same type, which themselves then interact with other neighbouring electrons, and thus a wave is formed in the sea of conductive (or nearly free) electrons. The little wave packet is called a *plasmon*, because it is an excitation of the *plasma* of conduction electrons.

Returning to the single electron interaction, we left an atom with a vacant position in its electronic structure. This is very rapidly filled by an electron from a higher electronic level falling into the vacant position. In doing this it must lose some energy (the difference between the two energy levels, in fact). It may do this in two ways: either by emitting an X-ray or by transferring the energy to a neighbouring bound electron which is emitted as an *Auger* electron. Auger electrons are mostly of such an energy that they only escape from the solid if they are created very close to its surface. As mentioned in Chapter 1, the surface chemical analysis technique of Auger Electron Spectroscopy is dealt with in another book in this series and will not be pursued further here.

The emitted X-rays have a wavelength which is absolutely characteristic of the atoms in which they were created. (They are called *characteristic* X-rays.) Thus monitoring the wavelengths of the X-rays emitted by a solid tells us immediately which elements are present in the solid. If we knew the relative efficiencies of production of the different X-rays by the various elements (i.e. the excitation *cross-sections*) we should be able to determine the chemical composition of the emitting volume. Unfortunately, between being created and arriving at the detector where they are counted, the X-rays may be absorbed by the rest of the solid through which they must travel. As a further complication, more X-rays of a given wavelength may be created by *fluorescence* by X-rays of higher energy. Thus the various physical processes involved in, or consequent on, inelastic collisions of the beam electrons with resident, solid state electrons, which we must understand before we are able to perform quantitative chemical microanalysis, are:

- the probability, or cross-section, for single electron excitation

- the relative probabilities of subsequent X-ray photon creation and of Auger electron creation

10 *Chemical Microanalysis Using Electron Beams*

- the probability of the X-ray being absorbed within the solid

- the probability of the X-ray intensity being augmented by fluorescence by X-rays of higher energy.

These possibilities are schematised in Fig. 2.3 and will be discussed in greater depth later in this chapter.

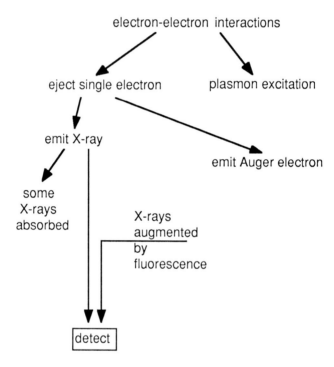

Fig. 2.3 The main physical stages in X-ray microanalysis.

I will now complete this overview of the interactions of the beam electrons with the solid by discussing much more briefly the other three relevant interactions. Although the heart of chemical microanalysis using electron beams is contained in the first section above on inelastic collisions with electrons, the other three types of interaction impinge on the analysis and must therefore be understood.

II. *Elastic scattering by atoms*. Here, the fast electrons in the beam interact with an atom taken as a whole. The interaction is mainly with the nucleus of the atom. The fast electron travels close to the nucleus and is attracted electrostatically by it, being scattered through a relatively large angle. This interaction is therefore important in determining the *trajectories* of the beam electrons, i.e. it controls *beam spreading*.

When the fast electron travels less close to the nucleus of the atom, it is scattered through a smaller angle and the orbital electrons begin to have some influence on the scattering.

III. *Inelastic scattering by nuclei*. For fewer than 1% of the nuclear scattering events described in the previous sub-section, II, a photon is emitted by the fast electron. This photon may have any energy between zero and the energy of the fast electron. The whole spectrum of radiation from this source is called *bremsstrahlung,* which is German for 'braking radiation'. This is important for chemical microanalysis because it forms a large proportion of the *background* on which sit the characteristic X-ray peaks, which are of primary interest to us.

IV. *Elastic scattering by the crystal.* If the atoms are regularly positioned in space to form a crystal lattice, then it is possible for the fast electrons to interact with the crystal as a whole. This is the familiar *Bragg scattering* which forms the basis of transmission electron microscopy of crystals. It is relatively peripheral to chemical microanalysis, although it will be mentioned in connection with microanalysis of thin foils in transmission.

It is worth noting that the terms *inelastic* and *elastic* are relative only. Scattering by single electrons is usually referred to as *inelastic* because it is so with respect to the atom, to which *internal* energy is transferred. It is equally fair to say, of course, that it is *elastic* with respect to the electron. Similarly, scattering by a single atom is elastic with reference to the atom, but inelastic with reference to the whole crystal.

The remainder of this chapter is devoted to a description of the individual interactions referred to above, which will enable us to interpret X-ray spectra quantitatively and thereby determine chemical compositions. The full list, and the appropriate section numbers, are as follows:

2.3 Single electron excitation
2.4 Plasmon excitation
2.5 X-ray and Auger electron emission
2.6 X-ray absorption
2.7 X-ray fluorescence
2.8 Bremsstrahlung
2.9 Elastic scattering by an atom
2.10 Bragg scattering
2.11 Summary

2.3 Single electron excitation

This is the most important (or at least the most central) of these individual interactions since it is the origin of the X-rays which are the basis of chemical microanalysis using electron beams. As a fast electron travels through a solid, it may be scattered by the solid as a whole, if it is crystalline (Bragg scattering), by atoms as a whole (diffuse elastic scattering), inelastically by nuclei ($\rightarrow$ bremsstrahlung) or inelastically by electrons. It is this last interaction which concerns us here. It is useful to separate inelastic scattering by electrons which are deep in the electronic structure of the atom (i.e. with low energy and close to the nucleus) from those which are either towards the outside of the atom, or actually free. In the former case the interaction of the beam electron is with a single (bound) electron. In the latter case the interaction is likely to be a collective interaction with the valence or free electrons in the solid. This section will deal with the first type of interaction, that with a single (bound) core electron. It is this type of interaction which ultimately leads to the X-rays used for chemical microanalysis.

The best place to begin a discussion of the interaction between the fast, beam electrons and a single bound electron in a solid is via the Rutherford scattering formulae.[1] (See also Appendix C.) These most useful of expressions describe in classical (i.e. non quantum mechanical) terms the scattering (trajectory etc.) of one charged particle as it moves past a second (target) charged particle. Of course, the trajectory of the target particle is also defined. In the present instance, both the target particle and the approaching particle are negatively charged electrons. The Rutherford scattering formulae will also be found useful in discussing the scattering of electrons by heavy, positively charged nuclei, in Section 2.9 on elastic scattering by atoms.

The Interactions of Electrons and X-rays with Solids

The Rutherford analysis tells us two important facts about any collision. It tells us about the *trajectory* of the incident particle and it tells us about its *loss of energy*. In this section it is the second aspect, the loss of energy, which is most important. In the section on elastic scattering by atoms, it is the change in direction of the incident particle which is of interest. Keep in mind, though, that it is the same analysis in both cases.

Whether it is the trajectory, or the loss of energy, which we wish to describe, the most convenient way to do it for present purposes is via *cross sections* and *differential cross-sections*. Thus the Rutherford analysis* describes the probability of the beam electron losing an energy E in a collision with a stationary electron in the following terms:

$$\frac{d\sigma}{dE} = \frac{\pi e^4}{(4\pi\varepsilon_0)^2 E_0} \tag{2.1}$$

$$\text{or } \sigma(E_0 > E > E_1) = \frac{\pi e^4}{(4\pi\varepsilon_0)^2 E_0}\left(\frac{1}{E_1} - \frac{1}{E_0}\right) \tag{2.2}$$

Here e is the electronic charge, E_0 is the original energy of the beam electron and ε_0 is the permeability of free space. I have left equations (2.1) and (2.2) in this form, with apparently redundant π's, to facilitate comparison with treatments based on cgs units, where the factor $(4\pi\varepsilon_0)$ does not appear. σ ($E_0 > E > E_1$) is the cross-section for scattering of the beam electron such that it loses an energy between some value E_1 and its entire kinetic energy E_0. (E_0 is obviously the maximum possible energy loss: it is only realisable in a collision between two particles of identical mass, such as here.) σ is an area. If you imagine the isolated target electron being irradiated by a uniform beam of incident electrons, any electrons which impinge on the area σ suffer an energy loss greater than E_1. σ is not a real, discrete, area, of course: it is describing the *probability* of some event - in this case inelastic scattering with an energy loss greater than E_1. If you think this is an awkward device to describe the probability of some event, try to think

---

*A good description of the Rutherford analysis may be found in Appendix B of 'The atomic nucleus' by R.D. Evans.[2]

of an alternative. You will soon be convinced of the beautiful simplicity of using cross-sections!

Equation (2.1) describes the variation of σ with E in a slightly different way. dσ/dE is a *differential* scattering cross-section. Equations (2.1) and (2.2) are exactly equivalent.

It is worth noting one or two features of equations (2.1) and (2.2). Firstly, the cross-section is inversely proportional to E_0. *The higher the energy of the incident electron, the less the chance of its losing a given amount of energy.* Thus, looking ahead a little, we will find that X-ray yields become smaller as the energy of the incident electron increases. The same inverse dependence of Rutherford cross-sections on E_0 also explains such apparently diverse phenomena as why polymers survive longer in a higher voltage microscope, why the beam penetration in high voltage microscopes is greater and why electron beam broadening in thin specimens becomes less serious at higher voltages.

Another interesting, but less propitious, aspect of equations (2.1) and (2.2) is that the cross-section becomes infinite as E tends to zero. These *soft* collisions (small E = soft; large E = hard) are due to fast electrons which pass a long way from the target electron. In fact, in real physical situations, either there are other charges in between the fast and target electrons, or, as below, the target electron is bound to a nucleus, and so this does not prove embarrassing.

The Rutherford cross-section (equations (2.1) and (2.2)) may be applied to the bound core electrons in a solid to give an expression for the ionisation cross-section for a particular quantum level. This was first effectively done by Bohr,[3] although when Bohr dealt with this problem, it was before the inception of the quantum theory and there was little idea of energy levels in atoms. Bohr was really interested in describing the overall energy loss of the fast electron as a function of distance.* If we consider, for example, the $1s^2$ or K shell bound electrons in a solid (see Appendix C), and apply equation (2.2), then the ionisation cross-section per atom will be:

---

*Like many 'classic' papers, those of Bohr[3] and Rutherford[1] are well worth reading in their original forms, where the clarity of their thinking is more apparent than in the increasingly distilled, if smoother, accounts found in successive generations of text books.

The Interactions of Electrons and X-rays with Solids

$$\sigma_K = \frac{z_K \pi e^4}{(4\pi\varepsilon_0)^2 E_0} \left(\frac{1}{E_K} - \frac{1}{E_0}\right)$$

where E_K is the ionisation, or binding, energy of a K shell electron and z_K (=2) is the number of electrons in the shell. Ignoring E_0^{-1} in comparison with E_K^{-1}, then

$$\sigma_K = \frac{\pi e^4 z_K}{(4\pi\varepsilon_0)^2 E_0 E_K}$$

In general, for the nl shell, where n is the principal, and l the orbital quantum number,

$$\sigma_{nl} = \frac{\pi e^4 z_{nl}}{(4\pi\varepsilon_0)^2 E_0 E_{nl}} \quad (2.3)$$

As far as excitation of core electrons is concerned, equation (2.3) has pushed the Rutherford analysis as far as (perhaps further than) it can go. Equation (2.3) is really a cheat, because I have virtually ignored the fact that the struck electron is bound (except in that I have chosen one of my integration limits to be E_{nl}). There is, however, a more fundamental objection to equation (2.3). For masses and distances on this scale, classical mechanics are inappropriate and quantum mechanics must be used. For a simple review of quantum mechanics, see, for example, Ref. 4. The important concepts to look up (for the purposes of this book) are the Schrödinger equation, the Dirac equation, the Heisenberg uncertainty principle and the Born approximation.

The most widely applied approach to this problem remains that of (the 24 year-old) Bethe, published in 1930.[5] The effect of the incident electron is considered as a perturbation on the struck atom. As with the Rutherford analysis, both energy loss and trajectory information can be extracted. At the moment, we are interested in energy losses. The total cross-section for removing an electron from the n,l quantum shell of an atom is[5]

$$\sigma_{nl} = \frac{\pi e^4 z_{nl}}{(4\pi\varepsilon_0)^2 E_0 E_{nl}} \, b_{nl} \, \ln\left(\frac{c_{nl} E_0}{E_{nl}}\right) \qquad (2.4)$$

It is comforting to note that the first of the two factors on the right hand side of equation (2.4) is identical to the right hand side of the Rutherford-inspired equation (2.3). The Rutherford cross section for free, unbound electrons is modified by an *inelastic form factor* which depends, via the dimensionless constants b_{nl} and c_{nl} in the formula above, on the details of atomic structure. Calculations of b_{nl} and c_{nl} of varying complexity can be performed, ranging from the *hydrogenic model* of Bethe (see also Chapter 5), wherein an atom with Z electrons is considered as an enlarged hydrogen atom with nucleus (= Z protons + (Z - 1) electrons) + 1 electron, up to Hartree-Slater models of the atom. (See Inokuti[6] or, for a shorter review, Inokuti and Manson.[7] Inokuti[6] presents in his section 4.4 an illuminating discussion of the relationship between Bethe, Rutherford and Bohr theories.)

Alternatively, an expression of the form of equation (2.4) can be fitted to experimental data. A plot of $(\sigma_{nl} E_0 E_{nl})/(\pi e^4 z_{nl})$ against $\ln(E_0/E_{nl})$ (the so-called *Fano plot*) will reveal over what range of E_0/E_{nl} the Bethe approach is valid (i.e. the straight line portion) and, over that range, the values of b_{nl} and c_{nl}. Fig. 2.4 shows an example of a Fano plot. E_0/E_{nl} is called - for obvious reasons - the *overpotential* and is written U_{nl}. The Bethe analysis is expected to (and does) fail for very low and very high U_{nl}. For low U_{nl} the *Born approximation*, on which the Bethe analysis depends, is invalid. The Born approximation requires that the speed of the beam electron should be much greater than the 'orbital speed' of the bound electron. Roughly (see Appendix C) this is $cZ/(137n)$ (c: velocity of light; Z: atomic number; n: principal quantum number) and thus $\beta = v/c$ for the beam electron should be much greater than $Z/(137n)$. Also, when the beam electrons are only just energetic enough to eject an electron from the nl shell, the ejection process becomes very sensitive to the details of the atomic structure. For example, the nl electron may not be excited to the continuum of unbound states, but may find itself arriving in a vacant, higher energy, bound state. At the other end of the scale, at high energies the Bethe approach may also fail, but this time because the nature of the collision changes, spin and exchange interactions become important.

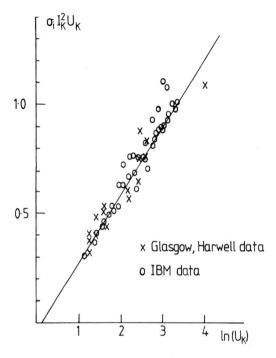

Fig. 2.4 An example of a Fano plot for a variety of elements at beam voltages ranging from 80 kV to 200 kV (Paterson et al.[8]). U_K, the overpotential, $= E_0/E_K$. I_K and σ_i are the E_K and σ_K of the present text. The two different plotting symbols refer to two different microscopes. The fact that data corresponding to several different elements fell on effectively the same straight line encouraged these authors to assign universal K shell Bethe parameters b_K and c_K (see text). (By kind permission of the *Journal of Microscopy*.)

Returning to Bethe cross-sections, the constants b and c in equation (2.4) have been fitted to a variety of expressions by a variety of authors. To some extent it depends on which energy range one is interested in. For bulk specimen analysis in SEMs the beam voltage is relatively low and so, therefore, are the overpotentials. A typical fit in this region is that of Green and Cosslett[9] who set c_{nl} to 1 for K shell excitation (see Appendix C) and b_K to 0.61. As another example, Powell,[10] by fitting a Bethe type expression to those experimental data which were available, found that for K shell

18 Chemical Microanalysis Using Electron Beams

electrons the Bethe approach, equation (2.4), was valid for $4 < U_K < 25$, with $b_K \sim 0.9$ and $c_K \sim 0.65$ for all elements. For the L_{23} shells,* for $4 < U_{L_{23}} < 20$, $b_{L_{23}}$ varied from ~ 0.6 (low Z) to ~ 0.9 (high Z) and $c_{L_{23}} \sim 0.6$. (These cross-sections also include the production of vacancies in the L shell by electrons falling from that shell to vacancies in the K shell to produce K X-rays (see below).) As a third and final variant, Paterson et al.[8] have recently given $b_K = 0.6$ and $c_K = 0.9$ (see Fig. 2.4), $b_L = 0.6$ and $c_L = 0.5$ (L_1, L_2 or L_3) based on their own experimental measurements. The reason that more accurate values of these very important parameters cannot be given is that they are rather difficult to measure, and one reason for *this* is that it is very difficult to disentangle cross-sections from X-ray detector efficiencies. Note that in a ratio, for example, between two different K shell peaks, b_K would cancel and c_K is in a logarithm and unlikely to affect the result very critically. As will be seen later, simple formulations like that of Powell enable standardless thin film microanalysis to yield compositions often within an accuracy of a few percent, which is frequently all that is required.

Equation 2.4 has the general form shown in Fig. 2.5. The maximum in the curve - the optimum overpotential - is at the point $U_{nl} = e/c_{nl}$, from equation (2.4). (Here e = the base of natural logarithms 2.718..., not the electronic charge.) Since c_{nl} will always be of the order of 1, the optimum overpotential will be ~ 3. Thus when an element giving rise to low energy X-rays is being analysed for, it may well be worth dropping the beam voltage to reduce U_{nl} (which is usually well above 3).

There is one final topic which I must deal with before embarking on some example calculations. The electrons in a transmission electron microscope fitted out for chemical microanalysis will typically be accelerated to potentials between 100 kV and 400 kV. At 100 kV the resultant speed of the electrons is 0.55 times that of light and at 400 kV it is 0.83 the speed of light. Thus for chemical analysis in transmission electron microscopes, relativistic effects** need to be taken into account. Equation (2.4) becomes (Bethe[12]):

$$\sigma_{nl} = \frac{\pi e^4 z_{nl}}{(4\pi\varepsilon_0)^2 E_0^{rel} E_{nl}} b_{nl} \left[\ln\left(\frac{c_{nl} E_0^{rel}}{E_{nl}}\right) - \ln(1 - \beta^2) - \beta^2 \right] \quad (2.5)$$

---

*The 2, 1 level, $= L_2 + L_3$ (see Appendix C), is the origin of the strongest L X-rays. Because L_2 and L_3 are so close in energy for medium Z they are often referred to collectively as L_{23}.

**A simple description of the relevant relativistic results may be found in Egerton,[11] pp 190-3.

where E_o^{rel} is the 'relativistically corrected' kinetic energy = $m_o v^2/2$; $\beta = v/c$; m_o is the rest mass of the electron; v is the speed of the electron and c is the velocity of light. Some further, more minor correction terms have been neglected. For an accelerating energy E, v may be written:

$$v = c \sqrt{1 - \frac{1}{\left(1 + \frac{E}{m_0 c^2}\right)^2}} \quad (2.6)$$

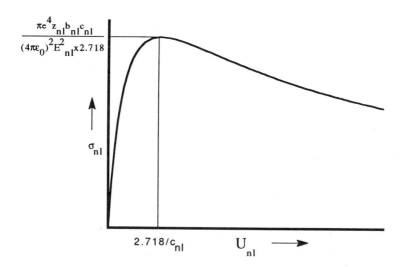

Fig. 2.5 The general form of the Bethe[5] cross-section (equation (2.4)) for ionisation of the nl electron shell.

Note that E in equation (2.6) is the *energy* of the electron, not its electrical potential. Beware of the difference between, for example, kV (an electrical potential) and keV (an energy, expressed in *kilo electron volts*). $m_o c^2$, the rest energy of an electron ~511 keV and so equation (2.6) may be rewritten:

20 Chemical Microanalysis Using Electron Beams

$$v = c\sqrt{1 - \frac{1}{\left(1 + \frac{E}{511}\right)^2}} \quad (2.6a)$$

where E is the beam energy expressed in keV.

Fitted parameters appropriate to equation (2.5) (note that the Fano plot must be modified to take into account the terms involving β) have been given by Zaluzec:[13]

$$b_K = 0.9880 - 0.01883\,Z + 3.0666 \times 10^{-4}\,Z^2 - 2.154 \times 10^{-6}\,Z^3$$

$$c_K = 0.2821 + 0.0770\,Z - 3.807 \times 10^{-3}\,Z^2 + 8.262 \times 10^{-5}\,Z^3 - 4.784 \times 10^{-7}\,Z^4 \quad (2.7)$$

which are stated to be valid between $U_K = 2$ and $E_0 = 1000$ keV.

A look at the remainder of the Zaluzec[13] paper will show why I have not included the L shell parameterisation. (See also the neighbouring articles in the same book by Egerton and by Newbury et al.)

Example calculation 2.1

What is the cross-section per atom for the ionisation of the K shell of aluminium by 25 kV electrons?

The relevant equation is (2.4):

$$\sigma_{nl} = \frac{\pi e^4\, z_{nl}}{(4\pi\varepsilon_0)^2\, E_0 E_{nl}}\, b_{nl}\, \ln\!\left(\frac{c_{nl}\,E_0}{E_{nl}}\right)$$

Here, n,l = 1, 0 = K shell. $z_K = 2$ (see Appendix C).

The Powell[10] values for b_K and c_K are 0.9 and 0.65 (see above). $E_0 = 25$ keV and from Appendix A, Table A.6, E_K for aluminium = 1.56 keV (to sufficient accuracy). Thus U_K, the overpotential, = 16 and we are within the range of validity of the Powell fitted constants (I would have used them in any case, probably!). Again from Appendix A, Table A.1,

The Interactions of Electrons and X-rays with Solids 21

$e = 1.6021 \times 10^{-19}$ C (i.e. $1eV = 1.6021 \times 10^{-19}$ J) and $\varepsilon_0 = 8.854 \times 10^{-12}$ F m^{-1} (i.e. C^2 J^{-1} m^{-1}).
Substituting into the formula above,

$$\sigma_K(\text{Fe}) = \frac{\pi (1.6021 \times 10^{-19})^4 \, 2 \times 0.9 \times \ln\left(0.65 \times \frac{25000}{1560}\right)}{(4\pi \times 8.854 \times 10^{-12})^2 (25000 \times 1.6021 \times 10^{-19})(1560 \times 1.6021 \times 10^{-19})}$$

$$= 7.0 \times 10^{-25} \text{ m}^2$$

N.B. Note that this cross-section is ~ 100 times smaller than the cross-sectional area of the (unscreened) Bohr $1s^2$ radius for aluminium (equation C.1, Appendix C). In general $\sigma_K \sim Z^{-2}$ of the area of the Bohr orbit.

Example calculation 2.2

The specimen in a TEM is a gold foil. At the point where the 120 kV beam passes through the foil, its thickness, parallel to the beam, is 100 nm. What proportion of the beam electrons eject L_{23} electrons from their bound states? (Gold has an fcc structure and a lattice parameter of 0.405 nm.)

First of all, we need the cross-section per atom for this ionisation. Again, we use equation (2.4):

$$\sigma_{nl} = \frac{\pi e^4 z_{nl}}{(4\pi\varepsilon_0)^2 E_0 E_{nl}} \, b_{nl} \, \ln\left(\frac{c_{nl} E_0}{E_{nl}}\right)$$

From Appendix A, Table A.1, $e = 1.6021 \times 10^{-19}$ C (i.e. $1eV = 1.6021 \times 10^{-19}$ J); $*_0 = 8.854 \times 10^{-10}$ F m^{-1} ($= C^2$ J^{-1}m^{-1}); $n,l = 2,1 = L_2 + L_3 = L_{23}$ shell. We use the Powell parameters: $b_{2,1} = b_{L_{23}} = 0.9$ (Au is heavy) and $c_{2,1} = 0.6$. $z_{2,1} = 6$. Again, from Appendix A, Table A.6, $E_{2,1}$ for Au $= 12.524$ keV (weighted average of E_{L_2} and E_{L_3}). $U_{2,1} = 9.6$ which is within the range specified by Powell.

22 Chemical Microanalysis Using Electron Beams

Substituting into the equation above for σ_{nl},

$$\sigma_{2,1} = \frac{\pi \left(1.6021 \times 10^{-19}\right)^4 6 \times 0.9 \times \ln\left(0.6 \times \frac{25000}{12524}\right)}{\left(4\pi \times 8.854 \times 10^{-12}\right)^2 \left(120000 \times 1.6021 \times 10^{-19}\right)\left(12524 \times 1.6021 \times 10^{-19}\right)}$$

$$= 4.0 \times 10^{-26} \text{ m}^2$$

For unit area of the foil (1 m² !) the volume is 100 nm x 1 m² = 10^{-7} m³. This volume contains $10^{-7}/(0.405 \times 10^{-9})^3$ unit cells, each of which contains four gold atoms. Therefore the total number of atoms / unit area in the specimen is 6.0×10^{21}. The total cross-section for L_{23} ionisation is $6.0 \times 10^{21} \times 4.0 \times 10^{-26} = 2.4 \times 10^{-4}$ m². Thus 0.024% of the beam electrons ionise L_{23} levels in the gold specimen.

Example calculation 2.3

What are the ratios of the Al(K) to Ni(K) cross-sections at 100 kV and 400 kV?

The high beam voltage 400 kV suggests we ought perhaps to use the relativistic version of equation (2.4) (i.e. equation (2.5)). Let us, however, use equation (2.4) first of all.

From Appendix A, Table A.1, e = 1.6021×10^{-19} C (i.e. 1eV = 1.6021×10^{-19} J); $\varepsilon_0 = 8.854 \times 10^{-12}$ F m^{-1} (= C² J^{-1}m^{-1}); nl = 1, 0 = K shell: the Powell[10] parameters are $b_K = 0.9$ and $c_K = 0.65$ and $z_K = 2$. From Appendix A, Table A.6, E_K for Al ~ 1.56 keV

and σ_K (Al, 100kV) =

$$\frac{\pi \left(1.6021 \times 10^{-19}\right)^4 \, 2 \times 0.9 \times \ln\left(0.65 \times \frac{100000}{1560}\right)}{\left(4\pi \times 8.854 \times 10^{-12}\right)^2 \left(100000 \times 1.6021 \times 10^{-19}\right)\left(1560 \times 1.6021 \times 10^{-19}\right)}$$

$$= 2.80 \times 10^{-25} \, m^2$$

Changing E_0 to 400 keV, σ_K (Al, 400 kV) = 9.61×10^{-26} m².
For Ni, E_K = 8.33 keV (Appendix A, Table A.6).
Then σ_K (Ni, 100 kV) = 2.89×10^{-26} m² and σ_K (Ni, 400 kV) = 1.21×10^{-26} m².

For the relativistic equation (2.5), we need $\beta = v/c$ and E_0^{rel} at the two beam voltages. Using equation 2.6a,

$$\frac{v}{c} = \sqrt{1 - \frac{1}{\left(1 + \frac{E}{511}\right)^2}}$$

where E is in keV, $\beta_{100\,kV}$ = 0.548 and $\beta_{400\,kV}$ = 0.828. $E_0^{rel} = m_0 v^2/2$. Remembering that $m_0 c^2$, the rest energy of the electron, is 511 keV, E_0^{rel} (100 kV) = $(1/2) \times 511 \times 0.548^2$ = 76.7 keV and E_0^{rel} (400 kV) = 175.2 keV (quite different from their nominal values!).

Using the Zaluzec[13] equations (2.7),

for Al (Z = 13), b_K = 0.79 and c_K = 0.81

and for Ni (Z = 28), b_K = 0.65 and c_K = 0.97

Recalling that equation (2.5) is

$$\sigma_{nl} = \frac{\pi e^4 z_{nl}}{(4\pi\varepsilon_0)^2 E_0^{rel} E_{nl}} b_{nl} \left[\ln\left(\frac{c_{nl} E_0^{rel}}{E_{nl}}\right) - \ln(1 - \beta^2) - \beta^2\right]$$

and substituting in all these values:

24 Chemical Microanalysis Using Electron Beams

$$\sigma_K (\text{Al}, 100 \text{ keV}) =$$

$$\frac{\pi \left(1.6021 \times 10^{-19}\right)^4 2 \times 0.79 \left\{ \ln\left(\frac{0.81 \times 76700}{1560}\right) - \ln\left(1 - 0.548^2\right) - 0.548^2 \right\}}{\left(4\pi \times 8.854 \times 10^{-12}\right)^2 \left(76700 \times 1.6021 \times 10^{-19}\right)\left(1560 \times 1.6021 \times 10^{-19}\right)}$$

$$= 8.60 \times 10^{-26} \left\{ \ln\left(\frac{0.81 \times 76700}{1560}\right) + 0.06 \right\}$$

$$= 8.60 \times 10^{-26} \left(3.68 + 0.06\right)$$

$$= 3.22 \times 10^{-25} \text{ m}^2$$

I have deliberately left the components in brackets separate so that you can see how small the relativistic correction is at 100 kV.

$$\sigma_K (\text{Al}, 400 \text{ keV}) =$$

$$3.77 \times 10^{-26} \left\{ \ln\left(\frac{0.8 \times 175200}{1560}\right) - \ln\left(1 - 0.828^2\right) - 0.828^2 \right\}$$

$$= 3.77 \times 10^{-26} \left\{ \ln\left(\frac{0.8 \times 175200}{1560}\right) + 0.47 \right\}$$

$$= 3.77 \times 10^{-26} \left(4.51 + 0.47\right)$$

$$= 1.88 \times 10^{-25} \text{ m}^2$$

- the relativistic correction is now appreciable.

The Interactions of Electrons and X-rays with Solids 25

and
$$\sigma_K \text{ (Ni, 100 kV)} = 3.00 \times 10^{-26} \text{ m}^2$$
$$\sigma_K \text{ (Ni, 400 kV)} = 2.02 \times 10^{-26} \text{ m}^2$$

Assembling these results into a table and working out the desired ratios:

	Non-relativistic (Powell[10])			Relativistic (Zaluzec[13])		
	100 kV	400 kV	100 kV / 400 kV	100 kV	400 kV	100 kV / 400 kV
Al	2.80×10^{-25}	9.61×10^{-26}	2.9	3.22×10^{-25}	1.88×10^{-25}	1.7
Ni	2.89×10^{-26}	1.21×10^{-26}	2.4	3.00×10^{-26}	2.04×10^{-26}	1.5
Al/Ni	9.7	7.9		10.7	9.2	

(All cross-sections in m².)

As will be seen in Chapter 4, at these high voltages, which are typical of TEMs, it is *ratios* of cross-sections which are important. Although the variations with voltage of the cross-sections are quite different for the two approaches (i.e. including $-\ln(1-\beta^2) - \beta^2$, or not) the variations of the ratios are not: $(Al/Ni)_{100 \text{ kV}} / (Al/Ni)_{400 \text{ kV}} = 1.22$ for the non-relativistic Powell numbers and 1.15 for the relativistic Zaluzec ones (incidentally 1.17 for non-relativistic Green-Cosslett[9] parameters: the experimental value (Z. Chen and M.H. Loretto (private communication) is 1.13).

In equation (2.4) we are interested in the cross-section for ionisation of one inner core energy level. The loss in energy of the beam electrons as they travel through a solid is controlled by frequent low energy collisions of the beam electrons with mainly outer, valence electrons belonging to the atoms in the solid. In the same paper where he derived equation (2.4), Bethe[5] also derived an expression for the average expected rate of energy loss of an electron beam as it passes through a solid:

$$-\frac{dE}{ds} = \frac{\pi e^4 Z}{(4\pi\varepsilon_0)^2 E} N^v 2 \ln\left(\frac{1.166 E}{J}\right) \quad (2.8)$$

26 *Chemical Microanalysis Using Electron Beams*

I have deliberately written equation (2.8) to mimic equation (2.4) to emphasise that it has the same sort of shape. In equation (2.8) s is path length, E is now a variable and N^v is the number of atoms per unit volume. J, which replaces E_{nl} in equation (2.4), is obviously some sort of statistical average for the atom and is called the 'mean excitation energy'. It was shown by Bloch[14] to be proportional, to a first approximation, to Z (roughly 11.5Z eV). In fact the constant 1.166 (= √(base of natural logarithms/2)), which is appropriate to electron beams, was supplied later by Bethe and Ashkin.[15] Note that (-dE/ds) will go through a maximum with respect to E just as σ_{nl} does. At relativistic voltages correction terms analogous to those of equation (2.5) should be added to the ln term. In this book, however, we will never be interested in stopping powers at TEM voltages (-dE/ds is too small to matter in transmission).

Later in this book I will occasionally use an earlier, less accurate, but simpler empirical expression for stopping power which originated in experimental measurements by Whiddington[16] and is usually called the Thomson-Whiddington law:[9]

$$-\frac{dE}{dz} = \frac{3.9 \times 10^{-25} \rho}{E} \quad Jm^{-1} \qquad (2.9)$$

where z is depth below the surface and ρ is measured in Mg m^{-3}. Note the similarity to the Bethe equation (2.8).

A closely related concept is that of the *absorption coefficient* for electrons. *Lenard's law* is usually written in terms of *mass absorption coefficient* thus:

$$\frac{I(z)}{I_0} = \exp(-\sigma\rho z) \qquad (2.10)$$

where σ is the mass absorption coefficient, called here the *Lenard coefficient*, and ρz is the *mass thickness*. (σ is being used here for a purpose totally different from that of a few pages ago, where it represented ionisation cross-section.) Mass absorption coefficients are preferred to absorption coefficients because any dependence on the local physical state of the absorber (e.g. density) is removed. Equation (2.10), like equation (2.9), is approximate. No account is taken of the energies of the electrons as they pass depth z - or emerge from the back of a thin foil of thickness z, although these two situations are different.

The Interactions of Electrons and X-rays with Solids 27

Note that, unlike equation (2.8), both equations (2.9) and (2.10) are written in terms of z (depth below the surface) and not s (path length). If it were required to calculate (2.9) and (2.10) accurately, not only equation (2.8) would have to be used, but some description of the high angle elastic scattering in the solid (see section 2.9).

Example calculation 2.4

What is the initial rate of energy loss of a 25 keV electron on entering a silicon specimen? What is the maximum rate (spatial, not temporal) of energy loss? At what electron energy does this occur? The (cubic) lattice parameter of silicon is 0.5431 nm and its unit cell contains 8 atoms.

The Bethe expression for energy loss is given above as:

$$-\frac{dE}{ds} = \frac{\pi e^4 Z}{(4\pi\varepsilon_0)^2 E} N^v 2 \ln\left(\frac{1.166 E}{J}\right)$$

N^v for silicon = $8 / (0.5431 \times 10^{-9})^3 = 4.994 \times 10^{28}$ atoms/m³.

Using data from Appendix A, Table A.1, and setting $J_{Si} = 11.5 \times 14$ (see text) = 161 eV,

$$-\frac{dE}{ds} =$$

$$\frac{\pi (1.6021 \times 10^{-19})^4 \, 14 \times 4.994 \times 10^{28} \times 2}{(4\pi \times 8.854 \times 10^{-12})^2 (25000 \times 1.6021 \times 10^{-19})} \ln\left(\frac{1.166 \times 25000}{161}\right)$$

$$= 3.035 \times 10^{-10} \text{ J m}^{-1}$$

$$= 1894 \text{ MeV m}^{-1} \quad \text{or} \quad \sim 1.9 \text{ eV nm}^{-1}$$

From equation (2.8) (-dE/ds) will be a maximum when E = eJ/1.166 (e here = base of natural logarithms) = 375.3 eV. If we still believe the Bethe formula (equation (2.8)) at this sort of energy, then

28 Chemical Microanalysis Using Electron Beams

$$-\frac{dE}{ds} =$$

$$\frac{\pi \left(1.6021 \times 10^{-19}\right)^4 \; 14 \times 4.994 \times 10^{28} \times 2}{\left(4\pi \times 8.854 \times 10^{-12}\right)^2 \left(375.3 \times 1.6021 \times 10^{-19}\right)} \ln\left(\frac{1.166 \times 375.3}{161}\right)$$

$$= 66 \; eV \; nm^{-1}$$

Actually, 375 eV is less than the K shell binding energy for silicon (see Appendix A, Table A.6) and equation 2.8 would obviously have to be adjusted if we were really interested in (-dE/ds) at these low electron energies.

This has been a long section, but as far as understanding the physical basis of electron beam chemical microanalysis is concerned, it is probably the most important in the book. We now go on to consider (much more briefly) the interaction between the fast beam electrons and the *outer* electrons of the atom - i.e. the valence and free electrons.

2.4 Plasmon excitation

When the struck electron is free, or nearly free, the interaction can often be described in terms of Rutherford scattering (see for example equation (2.1)). This is generally true when the beam electron passes close to the valence electron. When the beam electron passes at a greater distance from the valence electron, however, an alternative type of scattering may occur. The fast electron 'nudges' the valence electron which in turn nudges the neighbouring valence electrons, whose inertia resists this. Thus a wave is set up in the *plasma* of valence electrons. The interaction cannot be said to be with any particular electron in the solid : it is a *collective* excitation. The plasmon has its own preferred frequency of oscillation and only energies close to that corresponding to this resonance frequency are transferred from the fast electron. This energy quantised oscillation is called a *plasmon*; plasmon energies depend - *inter alia* - on valence electron densities and are typically of the order of a few eV. In fact, plasmons only survive for one or two wavelengths (of the order of 100 nm) before losing their energy to a single electron which itself transfers the energy to kinetic energy for an ion (heat) or to light in a process similar to that with which we shall shortly be dealing whereby X-rays are produced. Fig. 2.6 shows an energy loss spectrum for electrons transmitted through a foil. The peaks in the energy spectrum corresponding to plasmon creation may clearly be seen.

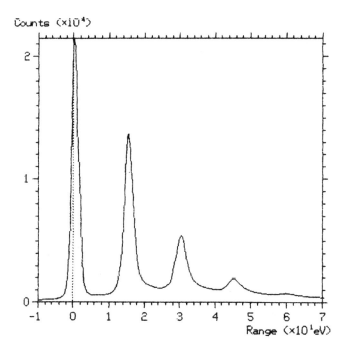

Fig. 2.6 An electron energy loss spectrum showing peaks due to plasmon losses. 'Counts' refers to numbers of electrons and 'Range' to electron energy loss. The specimen was aluminium - 3% magnesium. The plasmon energy is ~ 11 eV and the first three plasmon losses can be seen.

Plasmon production occurs very frequently and one might expect that it would have a large effect on the rate of energy loss of the fast beam electrons. In fact the effect is modest except at low beam electron energies. Humphreys[17] has shown that the right hand side of equation (2.8), which

30 *Chemical Microanalysis Using Electron Beams*

refers to single-electron excitations, must have added to it a term involving ln (J/E_p) where J is the mean excitation energy of an atom and E_p is the plasmon energy (usually of the order of 10 eV). This is equivalent to changing J to $J(J/E_p)^{-z_p/Z}$ where Z is the atomic number and z_p is roughly the number of 'valence' electrons. Provided empirically derived values of J are used (which will really be *effective* values, incorporating plasmon interactions) we may retain equation (2.8).

2.5 X-ray and Auger electron emission

In section 2.2 I described the inelastic scattering of the beam electrons by the bound electrons in the solid. When the scattering electron is a core electron, a vacancy is left in the electronic structure of the atom. This is rapidly filled by an electron from a higher (in energy terms) level. (This in its turn leaves a vacancy which is filled by a higher energy electron, etc., but these subsequent events need not concern us here.) As the electron drops from its original energy level to fill the vacancy created by the original electron collision, it must lose an amount of energy ΔE equal to the difference in energies between the two levels involved. This energy is either emitted as an X-ray photon whose frequency ν is determined by $h\nu = \Delta E$, or the energy ΔE is given to a third neighbouring bound electron which is then emitted from the atom. This is an *Auger* electron. (Note that in this case *two* vacancies are created in the atom.)

The first of the two mechanisms, in which an X-ray is emitted, is the origin of chemical microanalysis by X-rays. The energy of the X-ray emitted is uniquely characteristic of the element giving rise to it and indeed is called a *characteristic* X-ray. Not all transitions are allowed. For a strong X-ray line, the quantum number l must change by ± 1 and j must change by ± 1 or 0 (but cannot stay at 0) (see Appendix C). Physically, this corresponds to conservation of angular momentum, taking into account that of the X-ray photon.

The X-ray is designated by a letter corresponding to the principal shell of the ejected electron (i.e. K, L, M etc.) together with a Greek letter/Arabic number pair - for example K_α. Each Greek letter subscript refers to a group of X-rays of similar wavelength. The order of the letters follows no logic I have ever been able to discern, whereas the numbers relate to intensity within a group. Thus $L_{\beta 1}$ is the strongest L_β peak, or line. Table 2.1 shows the origins of the stronger lines. Note that the lower the atomic number of an element, the fewer lines it shows. Thus L_α, for example, involving as it

Line	Initial	Final
$K_{\alpha 1}$	K	L_3
$K_{\alpha 2}$	K	L_2
$K_{\beta 1}$	K	M_3
$K_{\beta 2}$	K	$N_{2,3}$
$K_{\beta 3}$	K	M_2
$L_{\alpha 1}$	L_3	M_5
$L_{\alpha 2}$	L_3	M_4
$L_{\beta 1}$	L_2	M_4
$L_{\beta 2}$	L_3	N_5
$L_{\beta 3}$	L_1	M_3
$L_{\beta 4}$	L_1	M_2
$L_{\beta 5}$	L_3	$O_{4,5}$
$L_{\gamma 1}$	L_2	N_4
$L_{\gamma 2}$	L_1	N_2
$L_{\gamma 3}$	L_1	N_3
L_{η}	L_2	M_1
L_{ι}	L_3	M_1
$M_{\alpha 1}$	M_5	N_7
$M_{\alpha 2}$	M_5	N_6
$M_{\beta 1}$	M_4	N_6
$M_{\gamma 1}$	M_3	N_5
$M_{\gamma 2}$	M_3	N_4
M_{δ}	M_2	N_4
M_{ϵ}	M_3	O_5

Table 2.1 Initial and final atomic states for the more intense characteristic X-ray lines (taken from Reed[18] by kind permission of Cambridge University Press).

32 Chemical Microanalysis Using Electron Beams

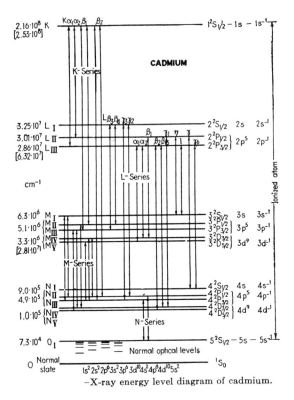

Fig. 2.7 Energy level diagram for cadmium taken from White[19] via Dyson[20]. (By kind permission of McGraw-Hill.)

does an electron dropping from M_5 to L_3, is absent for $Z < 23$ (V). Fig. 2.7 shows an example of an energy level diagram for cadmium.

The energy of any particular group of lines is proportional to $(Z - \text{constant})^2$, where the constant describes the effect of screening by the remaining electrons after the initial ionisation. (This is Moseley's law.) Fig. 2.8 demonstrates this for K series lines, where the screening constant is roughly 1 (generally 1 K electron remaining after ionisation). This Z dependence is consistent with equation (C.2) of Appendix C for binding energy.

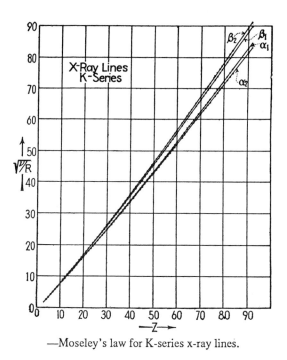

—Moseley's law for K-series x-ray lines.

Fig. 2.8 Moseley's law for K X-ray lines, taken from White[19] via Dyson.[20] (By kind permission of McGraw-Hill.)

The ratio

$$\frac{\text{X-ray photons emitted}}{(\text{X-ray photons} + \text{Auger electrons})} = \frac{\text{X-ray photons emitted}}{\text{number of original inelastic events}}$$

is known as the *fluorescent yield* for a transition. Since the Auger yield is largely independent of Z and the X-ray yield is proportional, via the Z^2 term in the binding energy of a shell, to Z^4, the fluorescent yield, ω, for a

34 Chemical Microanalysis Using Electron Beams

transition may be written approximately as $\omega = Z^4/(a + Z^4)$ where 'a' is a constant. To match experimental data more accurately, $Z/a^{1/4}$ is normally replaced by $(A + BZ + CZ^3)$, where A, B and C are fitted constants.

For example Bambynek et al.[21] give

$$\omega_K = \frac{f_K^4(Z)}{1 + f_K^4(Z)} \qquad (2.11)$$

where $f_K(Z) = 0.015 + 0.327Z - 0.64 \times 10^{-6} Z^3$

In general, however, the numbers are more safely looked up than generated, and specific values of ω, including L shells, are reproduced from Krause[22] in Appendix A, Table A.4. In the unlikely event of your needing to consult M shell fluorescence ratios, some are given by Bambynek et al.[21]

Thus by multiplying the cross-section for ionisation by the fluorescent yield, we can know how many K, L, etc. X-rays we may expect per incident electron. A glance at Fig. 2.7, however, shows that there are many component wavelengths in the K series, and even more so in the L, M, etc. series. Generally speaking, it is not convenient to measure the total intensity in a particular series. Rather, one measures the K_α intensity (= $K_{\alpha 1} + K_{\alpha 2}$), for example, or the L_α intensity. Therefore we need to know how the X-ray photons *partition* between the various lines. One might imagine this could be common sense. For example consider the α_1 and α_2 components of the K_α line. $K_{\alpha 1}$ arises from electrons falling from the L_3 shell to the K shell, $K_{\alpha 2}$ from $L_2 \rightarrow K$. Since there are 4 electrons in the L_3 shell and 2 in the L_2 (see Appendix C, Table C.1) the α_1 line is expected to be twice as strong as the α_2, which it often is. This simple approach, however, often fails for a number of reasons, of which two are:

1. As the energy levels comprising one set of lines become further apart, differences in ionisation energy and fluorescence efficiency begin to grow.

2. The original vacancy in the electronic structure caused by the beam electron may rearrange via a *Coster-Kronig transition* before an X-ray can be emitted. A Coster-Kronig transition is an intra-shell Auger transition. For example a vacancy in the L_1 shell might be filled by an electron from the L_3 shell, which ejects an Auger electron in the process. This would enhance those lines requiring an L_3 vacancy.*

The Interactions of Electrons and X-rays with Solids 35

It is therefore necessary to generate individual partition factors and suitable expressions[23] are reproduced in Appendix A, Table A.5. Some care is needed in using data from the literature: the method of producing the initial ionisation should be as nearly as possible the same.

Note finally that the strong groups of lines are, in fact, K_α and L_α.

Example calculation 2.5

What is the cross-section per atom for the production of Mn K_α X-rays by 20 kV electrons?

Using equation (2.4), the cross-section for K ionisation of Mn by 20 kV electrons (see Example calculations 2.1 and 2.2) =

$$\frac{\pi \left(1.6021 \times 10^{-19}\right)^4 2 \times 0.9 \times \ln\left(0.65 \times \frac{20000}{6537}\right)}{\left(4\pi \times 8.854 \times 10^{-12}\right)^2 \left(20000 \times 1.6021 \times 10^{-19}\right)\left(6537 \times 1.6021 \times 10^{-19}\right)}$$

$$= 6.2 \times 10^{-26} \text{ m}^2$$

The fluorescent yield ω_K from equation (2.11) is

$$\frac{f_K^4(Z)}{1 + f_K^4(Z)}$$

where $f_K(Z) = 0.015 + 0.327 \times 25 - 0.64 \times 10^{-6} \times 25^3$

$= 0.8225$

Thus $\omega_K = 0.31$, which compares with quoted individual experimental values in Table III. IV of Bambynek *et al.*[21] of 0.303 and 0.322 and in Table 3 of Krause[22] (see Appendix A, Table A.4) of 0.308.

The atomic cross-section for K X-ray production in Mn by 20 kV

electrons is therefore $6.2 \times 10^{-20} \times 0.31 = 1.9 \times 10^{-26}$ m². From Appendix A, Table A.5, $K_\beta / (K_\alpha + K_\beta)$ for Mn = 0.880 and the atomic cross-section for K_α X-ray production in Mn by 20 kV electrons is therefore
$1.9 \times 10^{-26} \times 0.880 = 1.7 \times 10^{-26}$ m².

2.6 X-ray absorption

We now know how many X-rays of a particular type are produced by electrons of a given energy. It would be nice to say at this stage that all we do is count them, thereby microanalysing quickly and conveniently a small volume of solid. Life, unfortunately, is not that simple. The characteristic X-rays that are produced are partially absorbed as they travel through the solid towards the X-ray detector. This absorptive process is the subject of this section.

As with the original discussion of electron interactions with solids at the beginning of this chapter, it is useful to classify the X-ray interactions with the solid according to whether they are with a single electron, with an atom or with the crystal. Again, it is useful to divide the single electron interactions into those with core electrons and those with valence electrons. This is shown in Fig. 2.9. The different interactions will be discussed briefly in turn. (See also Dyson,[20] Chapter 5.)

I. (Inelastic) interaction with single core electron

An X-ray (indeed any electromagnetic wave) consists of electric and magnetic fields oscillating at right angles to each other and to the direction of the ray. In an unpolarised X-ray, the electric and magnetic fields will oscillate in *all* directions perpendicular to the ray. The energy of the X-ray is quantised into packets of $h\nu$, where h = Planck's constant and ν is the frequency of the wave. This energy packet is associated with a *photon*. All of the interactions between an X-ray and a solid chiefly involve the electric field of the X-ray. When the photon energy $h\nu$ is less than the binding energies of the core electrons, the interaction tends to be with all the electrons together - i.e. the X-ray interacts with the atom as a whole. When the X-ray photon energy becomes greater than the binding energy of a particular electron, it may interact individually with it, ejecting it from the atom in a direction parallel to the electric field of the X-ray - i.e. normal to the direction of the X-ray. The ejected electron is called a *photoelectron*.

The Interactions of Electrons and X-rays with Solids 37

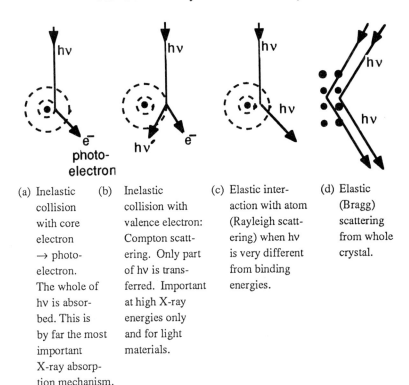

Fig. 2.9 The four interactions of an X-ray with a solid (compare Fig. 2.1 for electrons). (Don't take a particular line for a scattered X-ray too seriously - it is easier to draw a line than a spherical wave!)

The analysis of photoelectrons is the basis of XPS (X-ray Photoelectron Spectroscopy) which is an important surface analysis technique often available in the same instrument as Auger Electron Spectroscopy. The ejection of a core electron leaves a vacancy in the electronic structure which is filled by an electron falling from a higher level and from here on the sequence of events is the same as that described for electron irradiation in section 2.2. The characteristic X-rays which are given off are described as *fluorescence* of the solid under the irradiating X-ray beam. X-ray fluorescence (XRF) is an important analytical technique in its own right, but without the spatial resolution of electron beam techniques. X-ray fluorescence, as it enters electron beam microanalysis, will be discussed in the following section (2.7).

38 *Chemical Microanalysis Using Electron Beams*

This *photoelectron absorption* of X-rays is the single most important absorption mechanism. It is most efficient when the X-ray energy is just above the binding energy of the core electron, and is more important for the 'deeper' electrons, i.e. those nearer the nucleus, particularly the K electrons. I will delay a description of the cross-section for this process, until the other X-ray interactions with solids have been introduced (and dismissed).

II. (Inelastic) interaction with single valence electron

This is called *Compton* scattering. The X-ray loses *part* of its energy, in contrast to photoelectron absorption where the X-ray photon is either completely absorbed or not affected. This is because, in the photoelectron case, the momentum of the photon is transferred to the atom as a whole. For Compton scattering it is transferred to a single electron and more momentum would be required for complete energy transfer than the photon possesses. Compton scattering really only achieves importance as compared with photoelectron creation as the energy of the photon becomes of the order of the rest energy of the electron. At the energies typical of characteristic X-rays, the Compton effect is negligible. For 400 keV bremsstrahlung (see below) photons, however - particularly for a material of low density - it is the major absorption mechanism, albeit still very small: the cross-section is of the order of 10^{-28} m^2/electron.

III. (Elastic) scattering by atoms and
IV. (Elastic) scattering by crystals

When the X-ray photon energy is smaller than the binding energy of a core electron, it is scattered coherently by all such electrons in the atom. (The same is in fact true when the X-ray photon energy is much greater than the electron binding energy.) This type of scattering is called Rayleigh scattering. The cross-section is similar to that for Compton scattering. Because the effective scatterer is the atom, the energy loss is trivial. This is true even more so of that part of the Rayleigh scattering which is effectively from the crystal as a whole. This is the type of scattering which is used in X-ray diffraction, but for the purposes of this book it is unimportant.

Summary of absorption mechanisms

For characteristic X-rays the only important absorption mechanism is photoelectron absorption. Exact calculation of cross-sections is not easy for the same reason as for inelastic core electron ionisation cross-sections: the structure of the atom itself is an integral part of the calculation. As with the electron ionisation cross-sections, the usual approach is to fit a parameterised expression to experimental data, choosing the form of the fitted expression with half an eye on theoretically derived formulae.

Fig. 2.10 shows a typical curve of photoelectron excitation cross-section vs X-ray energy. The curve displays a series of discontinuities. Each of

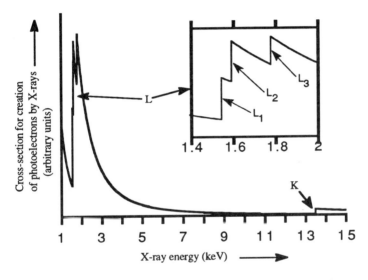

Fig. 2.10 The variation with X-ray energy of the photoelectron creation cross-section by X-rays (i.e. X-ray absorption cross-section) for bromine. The graph has been generated using the parameterisation of Thinh and Leroux[24] (see Appendix A, Table A.6). The discontinuities occur at the binding energies for various electronic shells, as indicated.

these corresponds to the ionisation of a particular energy level in the atom. K, L_1, etc., are marked. In fact it is not the cross-section which is normally tabulated, but the *mass absorption coefficient*, which has already been introduced in section 2.3 in connection with absorption of electrons. An absorption, or attenuation, coefficient μ is defined by

$$\frac{dI}{dx} = -\mu I \quad \text{or} \quad I \propto \exp(-\mu x) \quad \text{(Beer's law)} \quad (2.12)$$

where I is X-ray intensity and x is distance. Evidently $\mu = N^v \sigma$, where σ is the absorption cross-section per atom and N^v is the number of atoms per unit volume. The mass absorption coefficient is defined as μ/ρ, where ρ is the density of the absorbing medium. Recall that the reason for using μ/ρ rather than just μ is that this removes the dependence of μ on the local physical state of the absorber (section 2.3).

μ/ρ varies between *absorption edges* E_{nl} in the same way as shown in Fig. 2.10 for σ. In fact, (μ/ρ) varies roughly as $E^{-3.5}$ and as $Z^{4.5}$. For a simple explanation see Dyson,[20] p. 201 *et seq*. A convenient way to store and generate values of μ/ρ is to fit curves of the type:

$$\frac{\mu}{\rho} = CE_{nl}\lambda^{n_{nl}} \quad \text{or} \quad C_{nl}\lambda^{n'_{nl}} \quad (2.13)$$

to each section of the (μ/ρ) graph between the various absorption edges. This has been done by, for example, Thinh and Leroux[24] and their fitted parameters are reproduced in Appendix A, Table A.6. λ = wavelength in Å (= 12.3981 / E(keV)), (1Å = 0.1 nm), C is a constant for each element, E_{nl} (keV) is the n,l absorption edge at the low energy end of the interval containing the energy of interest and n_{nl}, n'_{nl} and C_{nl} are constants for each interval. The high energy range, above the K absorption edge, is divided for low Z atoms into two fitting regions by the energy E'. For this parameterisation, (μ/ρ) is in m² kg⁻¹ and E_{nl} is in keV. Alternative compilations are that of Heinrich[25] and, for low energy X-rays, that of Henke *et al*.[26]

Of course, most absorbing matrices contain mixtures of atoms. The attenuation coefficient is obtained by adding the cross-sections of all the various atom types:

$$\mu = N_1^v \sigma_1 + N_2^v \sigma_2 + \ldots \text{etc.}$$

$$= \sum N_i^v \sigma_i$$

where there are N_i^v 'i' atoms/unit volume, each with cross-section σ_i for absorbing the X-ray in question.

For the total absorption coefficient, then

$$\left(\frac{\mu}{\rho}\right) = \sum_i \left(\frac{\mu_i}{\rho_i}\right) C_i^m \qquad (2.14)$$

where C_i^m is the mass, or weight, concentration of element i.

Example calculation 2.6

To what fraction of its original intensity would a beam of copper K_α X-rays be reduced by 100 nm of Cu_3Au? (The lattice parameter of the $L1_2$ ordered intermetallic compound Cu_3Au is 0.374 nm.)

Remembering that absorption data are generated as mass absorption coefficients, Cu_3Au needs to be converted to wt %. From Appendix A, Table A.2, the relative atomic masses of copper and gold are 63.55 and 196.97 respectively. The weight percentage of gold in Cu_3Au is therefore 196.97/(3×63.55 + 196.97) = 50.8 wt %.

The energy of Cu K_α X-rays is 8.039 keV (Appendix A, Table A.3). Again from Appendix A, Table A.6,

For Cu at 8.039 keV, $\left(\frac{\mu}{\rho}\right) = 1.42775 \times 2.0961 \left(\frac{12.3981}{8.039}\right)^{2.73}$

$$= 5.11 \ m^2 kg^{-1}$$

For Au at 8.039 keV, $\left(\frac{\mu}{\rho}\right) = 1.94943 \times 3.4249 \left(\frac{12.3981}{8.039}\right)^{2.575}$

$$= 20.37 \ m^2 kg^{-1}$$

Thus for Cu_3Au at 8.039 keV, using equation (2.14),

$$\left(\frac{\mu}{\rho}\right)^{Cu_3Au} = 5.11 \times 0.492 + 20.37 \times 0.508 = 12.86 \ m^2 kg^{-1}$$

2.7 X-ray fluorescence

I return in this section to X-ray production by other X-rays. This was introduced previously (section 2.6) as one of the end results of X-ray absorption by photoelectron creation (Auger electrons being the other). I am interested here, however, not with absorption of the characteristic X-rays whose monitoring will enable us to estimate the concentration of the corresponding element. What I am interested in is *how many of these characteristic X-rays have been produced by the absorption of higher energy X-rays*. Clearly, if we are measuring and comparing intensities of characteristic X-rays, it is important to know what proportion of them originates directly from the electron beam, and what proportion is fluoresced by higher energy X-rays. The full answer to this question will be supplied later, in sections 4.1.2 (thin specimens) and 4.2.2.4 (bulk specimens). Here I will just mention some preliminary and qualitative aspects. It is also perhaps pertinent at this stage to look slightly ahead (section 2.8) and note that not only are the higher energy X-rays characteristic ones: there is also a continuous (in energy terms) background of X-rays which fluoresce other lower energy X-rays.

Everything we will need to know about the fluorescence process is implicit in the absorption data given in the last section (2.6). Fig. 2.11 is an enlargement of the K absorption edge in Fig. 2.10. Evidently the fraction of the mass absorption coefficient connected specifically with ionisation of the K electrons is AB/AC in the diagram, or, in terms of the absorption edge jump ratio 'r', $(r - 1)/r$. (r is defined as AC/BC.) Noting from the table of absorption coefficients in Appendix A, Table A.6, that the exponents n_{nl} in the mass absorption parameterisation expression are fairly constant for the same element, the ratio AB/AC will not vary very much with the energy of the fluorescing X-ray. Thus, whatever matrix the fluoresced atom finds itself in, AB/AC of the absorption of any wavelength X-rays by this element will correspond to K shell ionisation. Similar logic applies to the other absorption edges.

Example calculation 2.7

What fraction of calcium atom ionisations caused by X-rays with energy >4.0381 keV corresponds to K shell ionisations?

The Interactions of Electrons and X-rays with Solids 43

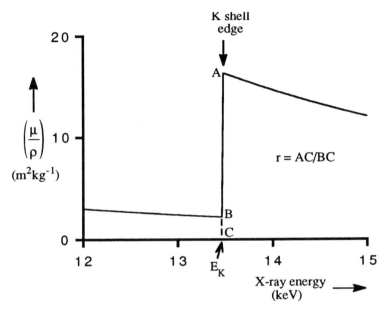

Fig. 2.11 The K absorption edge of bromine (magnified from Fig. 2.10). The mass absorption coefficient (μ/ρ) plotted here is proportional to the photoelectron absorption cross-section plotted in Fig. 2.10. The fraction of the absorption coefficient on the high energy side of E_K which is due to K shell ionisation is obviously AB/AC. The absorption edge jump ratio 'r' is defined as AC/BC.

From Appendix A, Table A.3, the K absorption edge for calcium is at 4.0381 keV. Using the data from Appendix A, Table A.6, appropriate to the high energy side of the edge,

$$\left(\frac{\mu}{\rho}\right)_{\text{high E}} = 4.919 \left(\frac{12.3981}{4.0381}\right)^{2.7345}$$

$$= 105.7 \text{ m}^2\text{kg}^{-1}$$

From the low energy side,

$$\left(\frac{\mu}{\rho}\right)_{\text{low E}} = 1.22904 \times 0.4378 \left(\frac{12.3981}{4.0381}\right)^{2.73}$$

$$= 11.5 \text{ m}^2\text{kg}^{-1}$$

Therefore the fraction corresponding to K shell ionisations

$$= \frac{105.7 - 11.5}{105.7} = 0.891$$

You may recall that it is helpful to assume that this fraction is unaffected by voltage. Let us check this roughly by extrapolating the low energy side data in the calculation above out of their legitimate energy range. For example, at 10 keV,

$$\left(\frac{\mu}{\rho}\right)_{\substack{\text{'high E' - i.e. including} \\ \text{K ionisations}}} = 4.919 \left(\frac{12.3981}{10}\right)^{2.7345}$$

$$= 8.854 \text{ m}^2\text{kg}^{-1}$$

$$\left(\frac{\mu}{\rho}\right)_{\substack{\text{'low' E - i.e. excluding} \\ \text{K ionisations}}} = 1.22904 \times 0.4378 \left(\frac{12.3981}{10}\right)^{2.73}$$

$$= 0.968 \text{ m}^2\text{kg}^{-1}$$

The fraction corresponding to K shell ionisations

$$= \frac{8.854 - 0.968}{8.854} = 0.891$$

Taking into account also that the absorption coefficient has dropped by more than a factor of 10, the approximation of ignoring changes with energy in the fraction of (μ/ρ) corresponding to some particular edge (i.e. changes in the n_{nl}) is obviously a good one.

2.8 Inelastic collisions between beam electrons and nuclei : bremsstrahlung

Thus far in this chapter we have been concerned exclusively with inelastic collisions between the fast beam electrons and electrons in the solid to be analysed, and the results of this interaction. At this point I wish to move on to a totally different type of interaction, introduced in section 2.2 (III): an inelastic interaction between the beam electron and the nucleus, in which an X-ray photon is emitted. These X-rays are called *bremsstrahlung* and form a continuous background to the characteristic peaks.

Bremsstrahlung is emitted when the beam electrons pass close to a nucleus and are scattered through large angles. Classical electrodynamics dictates that when a charge (= fast electron) is accelerated (scattered through large angle → velocity changed quickly → (centripetal) acceleration) it will radiate. Such a theory was developed by Kramers.[27] More accurate quantum mechanical theories were subsequently introduced by Sommerfeld[28] (non-relativistic) and Bethe and Heitler[29] (relativistic). The main physical difference between the classical and quantum theories is that in the former, *every* close encounter of the fast electron with a nucleus results in bremsstrahlung whereas the latter predict that only a small proportion of the collisions - roughly 1/137 - do so (although with a compensatorily high yield per event). The effect of nuclear scattering on the directions of the electrons, including the vast majority of cases where no bremsstrahlung is emitted, will be described in the next section, 2.9.

The main results of theory and experiment are as follows:

(a) *Energy dependence* : following Kramers,[27] the energy differential cross-section per atom (see Section 2.2) for bremsstrahlung production (measured in photons*)

$$\frac{d\sigma_{br}}{dE} \propto \frac{Z^2}{m_0 E E_0} \qquad (2.15)$$

where E_0 is the energy of the fast electrons, E is the bremsstrahlung energy, Z is the atomic number of the target and m_0 is the rest mass of the electron.

---

*Often given in terms of *energy flow*: flux of photons x energy per photon (hν).

46 *Chemical Microanalysis Using Electron Beams*

The quantum mechanical theories modify this relationship by introducing a factor B, where B is a slowly varying function of Z and E_o.

It is not possible to compare directly equation (2.15) with equation (2.4) for ionisation because they are chalk and cheese. Equation (2.4) is a total cross-section, integrated over energy, whereas equation (2.15) is a differential cross-section. How the characteristic radiation described by equation (2.4) (which we will wish to measure) compares with the background defined by equation (2.12) will depend very much on what energy range the characteristic X-rays are spread over. This in its turn will depend on basic physical limitations, like how long the electron takes to drop into the low energy level (through the Heisenberg uncertainty relationship linking time with energy) and on the energy resolution of the detector we use. What we can say, however, from equations (2.15) and (2.4) is that the ratio of characteristic X-rays to bremsstrahlung (background) should (all other things being equal)

- increase with beam energy E_o
- decrease with atomic number Z
- decrease with m_o

In fact, what we actually measure is dominated by other factors - for example the loss in energy of the beam as it penetrates into a bulk specimen, or electron scattering within the instrument.

The reason for including m_o, the rest mass of the electron, which is totally immutable of course, is that this expression works equally well for protons and the low background to characteristic X-ray peaks from proton irradiation is the attraction of PIXE (Proton Induced X-ray Excitation). (Keep in mind, though, that a high energy proton is much more expensive than a high energy electron, and much more difficult to control (focus)!)

(b) *Angular dependence* : unlike characteristic X-rays, bremsstrahlung is emitted in a very anisotropic manner. With characteristic X-rays, the emission of the X-ray is a separate event from the initial ionisation and, as such, there is no reason for it to be in any particular direction. In the case of bremsstrahlung, however, the emission of the X-ray photon happens at the same time as the electron is scattered, and is very much tied to the electron's direction. The angular dependence is important mainly in transmission. The rules for bremsstrahlung production from thin films are these:

(i) As the beam voltage rises the direction of maximum bremsstrahlung intensity moves from lying at right angles to the beam to lying parallel with it. (This influences the positioning of X-ray detectors in TEMs.)

(ii) For the same beam voltage the lower energy bremsstrahlung is peaked in a forward direction to a greater extent than the high energy bremsstrahlung.

(iii) Z has little effect on the angular distribution of bremsstrahlung (but a considerable effect on its overall intensity - see above).

These trends are illustrated in Fig. 2.12. Keep in mind that, as with the energy dependence of bremsstrahlung discussed above, actual measurements tend to deviate rather widely from these predictions.

2.9 Elastic scattering of the beam electrons by atoms

This was introduced in section 2.2 (II). The high angle scattering is dominated by the effect of the nucleus on the fast electrons. For moderate electron energies this is described very well by the Rutherford analysis. The last time I used this (section 2.3) was for electron scattering by individual electrons. Now we are interested in electron scattering by a massive, positively charged nucleus. The same analysis (see again Evans,[2] Appendix B) yields, of course, rather different formulae for this very different situation. We are concerned here with angle, not energy, and thus the energy differential cross-section used earlier is replaced by an angle differential cross-section. The geometry is shown in Fig. 2.13. The increment in cross-section corresponding to scattering into a conical element of solid angle $d\Omega$, itself corresponding to scattering at an angle θ, is written $d\sigma/d\Omega$

and
$$\frac{d\sigma}{d\Omega} = \left(\frac{Z e^2}{8\pi\varepsilon_0 m_0 c^2} \frac{\sqrt{1-\beta^2}}{\beta^2} \frac{1}{\sin^2\frac{\theta}{2}} \right)^2 \qquad (2.16a)$$

48 *Chemical Microanalysis Using Electron Beams*

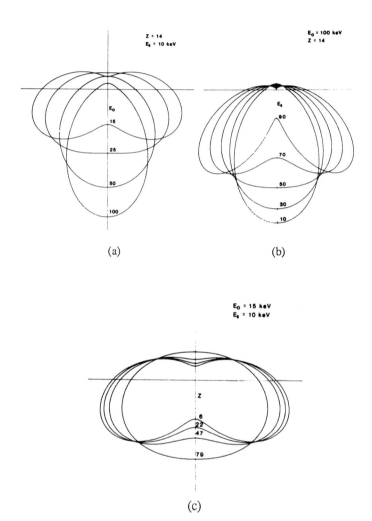

Fig. 2.12 Bremsstrahlung production from thin films, demonstrating the effect of (a) beam energy ($Z = 14$, $E_{brem} = 10$ keV) (b) bremsstrahlung energy ($Z = 14$, $E_o = 100$ keV) and (c) atomic number ($E_o = 15$ keV, $E_{brem} = 10$ keV). The areas of the plots have been normalised to facilitate comparison. E_{brem} is written E_b in the figures. (Taken from Russ[30] by kind permission of Butterworths.)

The Interactions of Electrons and X-rays with Solids 49

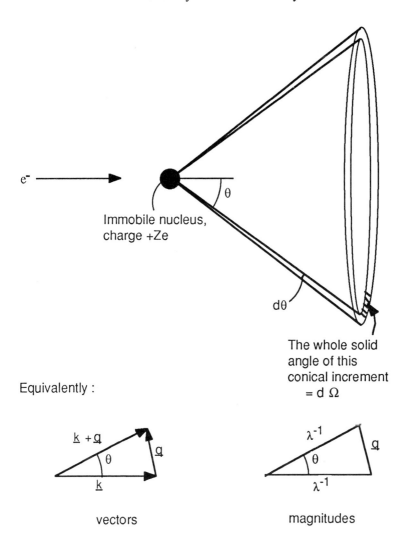

Fig. 2.13 The geometry of Rutherford scattering.

Equivalently,

$$\frac{d\sigma}{d\Omega} = \left(\frac{Z}{2\pi^2 \sqrt{1-\beta^2}\, a_H\, q^2}\right)^2 \quad (2.16b)$$

where $\beta = v/c$ with v the velocity of the electron, a_H is the Bohr radius (see Appendix C), Z is the atomic number = the charge on the nucleus and q is the magnitude of the *scattering vector* $\underline{q}$ and is defined in Fig. 2.13. Notice that q has the dimensions (1/length) and is said to exist in *reciprocal space*, which you may be familiar with from electron microscopy. Sometimes the 'q' form of the equation is more convenient.

Equations (2.16) are derived from classical mechanics and include all collisions of the electron with the nucleus, whether bremsstrahlung-emitting or not. In classical theory, *all* collisions result in bremsstrahlung.

The square root of the angle differential scattering cross-section is called the *scattering factor* and is written $f(\theta)$ or $f(q)$. The scattering factor is a frequently used concept. It should be noted that it is a *derived* quantity and does not correspond to a real measurable quantity, unlike the differential cross-section. Consistent with this is the fact that the scattering factor can evidently be complex (in the mathematical sense). The imaginary part of the scattering factor may correspond physically either to a phase factor or to absorption.

Equations (2.16) are consistent with the Schrödinger equation. They are also relativistically correct, but neglect the effects of spin. These are included in the Dirac equation, use of which results in equations (2.16a) (say) being multiplied by a correction factor (Mott [31]).

The correction factor is not expressible in simple analytical form. It depends on Z, E and θ. Thus:

1. As $\theta \to 0$, the correction factor $\to 1$ (i.e. the relativistic Rutherford expression is correct) for all combinations of Z and E.

2. For light atoms (small Z) the Rutherford expression is correct for small electron energies. At higher energies there is an approximate expression for the correction factor which, in the version due to McKinley and Feshbach,[32] alters equation (2.16a) as follows:

$$\frac{d\sigma}{d\Omega} =$$

$$\underbrace{\left(\frac{Z}{2\pi^2 \sqrt{1-\beta^2}\, a_H\, q^2}\right)^2}_{\text{Rutherford}} \times \underbrace{\left(1 - \left(\beta \sin\frac{\theta}{2}\right)^2 + \pi \beta \frac{Z}{137}\left(1 - \sin\frac{\theta}{2}\right)\sin\frac{\theta}{2}\right)}_{\text{Mott}}$$

...(2.17)

The $(\beta \sin \theta/2)^2$ term in the correction is the more important part. The last term is often omitted. Note that as θ or $\beta \to 0$, the correction $\to 1$.

3. For heavy atoms the Rutherford expression is invalid even at low electron energies. There is no approximation of simplicity comparable to equation (2.17), although McKinley and Feshbach[31] give a series expansion in terms of $Z/137$ and $Z/(137\beta)$ (note the β on the bottom line: low electron energies) involving coefficients which need to be calculated numerically. In fact the original expressions are trivial to evaluate in the current computer age.

It is significant that the Rutherford formula is not appropriate for heavy atoms at low beam voltages. This means that beam spreading calculations based on the Rutherford formula (which they nearly always are) are not only incorrect at high TEM beam voltages (e.g. 400 kV) but also for heavy materials for SEM type beam voltages. How 'low' and 'high' are to be interpreted should be elucidated from the original references. (I suggest you start with Mott and Massey.[33]) The trends described above are illustrated in Fig. 2.14. The sometimes large discrepancies between Rutherford and Mott calculations are partly because the Mott approach includes more that is new than spin alone.[34]

Returning to the general discussion of elastic scattering of electrons by atoms, large angles of scattering correspond to the electron's passing close to the nucleus. If the trajectory of the electron lies further away from the nucleus the angle of scattering decreases and the electron becomes more and more influenced by the orbital electrons of the atom. Thus for small angle scattering, Z in formula (2.16) has to be replaced by $Z - f_x(\theta)$, where $f_x(\theta)$ is the atomic scattering factor for X-rays. The calculation of $f_x(\theta)$ is not trivial, requiring, like that of b_{nl} and c_{nl} in equation (2.4), a detailed knowledge of atomic structure. In the calculation of *extinction distances* (a closely related parameter) for electron microscopy, this is done as accu-

52 Chemical Microanalysis Using Electron Beams

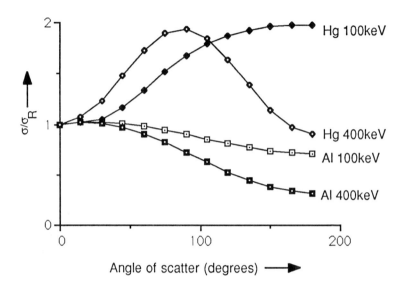

Fig. 2.14 The ratio of Mott to Rutherford cross-sections (σ/σ_R) for nuclear scattering of electrons. (Based on results given by Doggett and Spencer.[35])

rately as possible. For our purposes here, where we are interested only in the high angle scattering, equation (2.16) can be used as such, or with q^2 in equation (2.16b) replaced by $q^2 + q_0^2$, where q_0 describes the *screening* of the nucleus by the intervening orbital electrons. Another useful approximation is to replace Z by $\sqrt{Z(Z+1)}$ to describe approximately the additional effect of scattering by single electrons, discussed earlier in the chapter (Lenz[36]).

Example calculation 2.8

What proportion of a 100 kV electron beam is scattered by more than 10° by 200 nm of silicon? What proportion is backscattered? (The diamond cubic lattice parameter of silicon is 0.5431nm.)

Notice first of all that the Rutherford cross-section becomes infinite as

$\theta \to 0$. This is because, in principle, any electron, wherever it is in the universe, will be affected by our totally unscreened nucleus. Provided scattering angles corresponding to realistic *impact parameters* (nearest distance to nucleus, ignoring scattering) are used there is, however, no problem. For scattering by an angle greater than θ, the cross-section will be (see Fig. 2.13):

$$\int_\beta^\pi \frac{d\sigma}{d\Omega} d\Omega = \int_\beta^\pi \frac{d\sigma}{d\Omega} 2\pi \sin\theta \, d\theta$$

Substituting from equation (2.16a),

$$\sigma(>\theta) = \frac{Z^2 e^4 (1-\beta^2)}{32 \pi \varepsilon_0^2 m_0^2 c^4 \beta^4} \int_\theta^\pi \frac{\sin\theta}{\sin^4 \frac{\theta}{2}} d\theta$$

$$= \frac{Z^2 e^4 (1-\beta^2)}{16 \pi \varepsilon_0^2 m_0^2 c^4 \beta^4} \cot^2 \frac{\theta}{2} \qquad (2.18)$$

At 100 kV, $\beta = 0.548$ (equation (2.6a)). For silicon $Z = 14$. Using various data from Appendix A, Table A.1

$$\sigma(>\theta) = 3.79 \times 10^{-26} \cot^2 \frac{\theta}{2} \quad m^2 (\text{atom of Si})^{-1}$$

For $\theta = 10°$, $\sigma = 4.96 \times 10^{-24}$ m^2 atom^{-1}
For $\theta = 90°$ (i.e. backscattering), $\sigma = 3.79 \times 10^{-26}$ m^2 atom^{-1}

Imagine an area A m^2 of foil. Its volume will be A.(200×10^{-9}) m^3. Each silicon unit cell contains 8 atoms and has a volume of 0.5431^3 nm^3. 'A' nm^2 of foil therefore contain $A \times 200 \times 10^{-9} \times 8/(0.5431 \times 10^{-9})^3$ silicon atoms with a total cross-section for scattering at angles greater than 10° of $A \times 200 \times 10^{-9} \times 8 \times 4.96 \times 10^{-24}/(0.5431 \times 10^{-9})^3$ m^2. The fraction of electrons scattered by more than 10° (at least in single events) = $200 \times 10^{-9} \times 8 \times 4.96 \times 10^{-24}/(0.5431 \times 10^{-9})^3 = 0.0496$. Similarly the fraction backscattered = 0.04%.

Example calculation 2.9

What is the ratio between the Rutherford and Mott elastic scattering cross-sections for copper at 400 kV and 30°? 180°?

From equation 2.6a, at 400 kV

$$\frac{v}{c} = \sqrt{1 - \frac{1}{\left(1 + \frac{E}{511}\right)^2}}$$

$$= 0.828$$

I will use the approximate expression of equation (2.17). The ratio of the Mott to Rutherford cross-sections is the factor to the right within parentheses:

$$1 - \left(\beta \sin \frac{\theta}{2}\right)^2 + \pi \beta \frac{Z}{137}\left(1 - \sin \frac{\theta}{2}\right) \sin \frac{\theta}{2}$$

$$= 1 - (0.828 \times 0.259)^2 + \pi \, 0.828 \, \frac{29}{137} (1 - 0.259) \, 0.259$$

$$= 1.06$$

At 180° (i.e. for backscattering) this ratio is independent of Z (in this approximation) and $= 1 - \beta^2 = 0.314$ at 400 kV.
This is a substantial correction!

2.10 Elastic scattering of the beam electrons by crystals

In this section we come to the end of our journey through the various interactions between fast electrons and solids. It will just remain in section 2.11 to summarise the main results of this chapter.

For those solids which are crystalline it is possible for the beam electrons

The Interactions of Electrons and X-rays with Solids 55

to interact with the crystal as a whole. This is what is usually referred to as Bragg diffraction. Since the interaction is with the crystal as a whole, the energy loss by the beam electrons is many orders of magnitude less than the already effectively negligible loss in single atom elastic scattering. Although Bragg diffraction is behind all the enormous subject of the electron microscopy of crystalline materials, it is only important here for one fairly restricted reason: when electrons are scattered strongly by a set of crystal planes, their spatial distribution changes (Fig. 2.15). When the crystal is set so that the deviation parameter 's' is negative,* the electrons are channelled

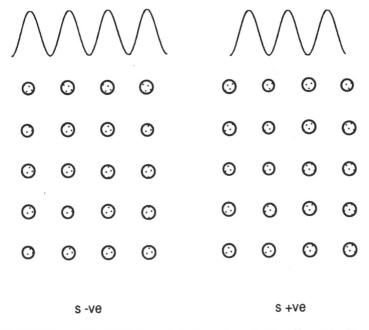

Fig. 2.15 The spatial distribution of electrons is markedly affected by Bragg (crystal) scattering. For s-ve (see text) the electrons are channelled along the atoms and the electron distribution tends towards that depicted in the left hand picture. For s+ve the electrons are channelled between the atoms (right hand picture). (Adapted from Hirsch et al.[37] by kind permission of Butterworths.)

---

*The incident electron beam makes an angle with the diffracting crystal planes which is smaller than the exact Bragg angle $\sin^{-1}(\lambda/2d)$; d = plane spacing.

along the atom planes. When 's' is positive, the electrons are channelled *between* the atoms. This means that for negative 's', say, processes localised to the atom, like X-ray production or single atom elastic scattering, will be enhanced, and *vice versa*. This can sometimes be turned to good account, as will be seen in a later chapter.

2.11 Summary of Chapter 2

When a solid is irradiated by a beam of fast electrons, some of the core electrons are ejected. The cross-section for this is described by equations (2.4) (SEM voltages) or (2.5) (TEM voltages) with the relevant parameters defined in the text or in equations (2.7). The filling of a fraction of the resulting orbital electron vacancies results in the emission of a characteristic X-ray. The relevant fluorescence yield is described by equations (2.11). Probably only some of the appropriate group of X-rays (e.g. K, L, etc.) are measured. The partition factor may be calculated via data presented in Appendix A, Table A.5. The X-ray may be absorbed on its way to the detector. Parameterised mass absorption coefficients will also be found in Appendix A, Table A.6. Higher energy X-rays may augment the measured X-rays by fluorescence, as described by the edge jump data implicit in the mass absorption coefficients. The characteristic peaks reside on a background of which a substantial part is formed by bremsstrahlung.

This summary describes most of the important events in thin foil microanalysis in a TEM. In bulk specimen microanalysis in an SEM we will also have to worry about the volume of specimen irradiated by the electrons and the way in which the electrons slow down. This is determined by the energy losses of the electrons (equation (2.8)) and their (single atom) elastic scattering (equations (2.16a,b)). The physics developed in this chapter will be applied to the specific situations met with in SEM and TEM microanalysis in Chapter 4. Before that, however, it is necessary to discuss the instruments themselves and the X-ray detectors. This will be done in Chapter 3.

Chapter 3 Microscopes and Spectrometers

In this Chapter I shall describe three types of electron microscope and two types of X-ray spectrometer. The three microscopes are the Scanning Electron Microscope (SEM) / Electron Probe Microanalyser (EPMA) used for bulk specimens, the Transmission Electron Microscope (TEM) used with thin specimens and the hybrid dedicated Scanning Transmission Electron Microscope (STEM). The two X-ray spectrometers are the Energy Dispersive X-ray Spectrometer (EDX or EDS) and the Wavelength Dispersive X-ray Spectrometer (WDX or WDS).

3.1   Scanning electron microscope / electron probe microanalyser

I mentioned in Chapter 1 that SEMs and EPMAs are growing closer and I shall make no very sustained effort to distinguish between them. From this point on, 'SEM' means 'SEM or EPMA', unless an explicit distinction is drawn.

Fig. 3.1(a) shows a photograph of an SEM. Fig. 3.1(b) is a schematic diagram of the interior of the microscope and Fig. 3.1(c) illustrates rather more clearly than does Fig. 3.1(b) the beam electron trajectories in the conventional mode of operation. In this mode a small electron probe is scanned in a rectangular raster across the surface of a bulk specimen. One of the resulting signals (e.g. the 'secondary' electron intensity - see below) is displayed synchronously on a cathode ray tube (CRT).

At the top of the microscope (in this example) is the source of electrons called the *electron gun*. There are three common types of gun, incorporating:

    (a)   a heated tungsten filament,
    (b)   a heated LaB_6 filament, or
    (c)   a tungsten field emission tip.

Passing through the sequence from (a) to (c) corresponds to:

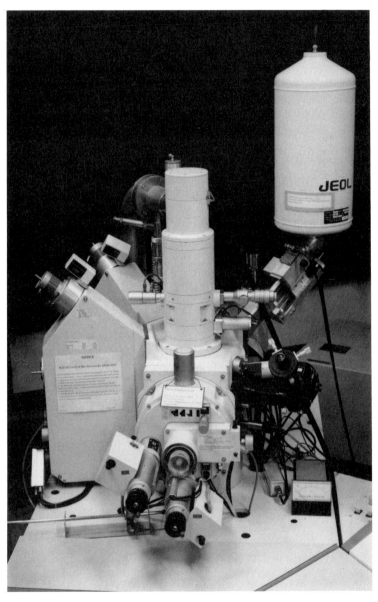

Fig. 3.1 (a) A scanning electron microscope (SEM) (JEOL 840A).

(i) less common,
(ii) more expensive,
(iii) better vacuum required
and (iv) smaller probe for the same current.

The attraction is (iv) on the list. Field emission guns are essential for dedicated scanning *transmission* microscopes (STEMs).

Moving down the microscope 'column', there are two *condenser* lenses and then two sets of *scanning coils* whose combined function is to move the probe across the specimen surface. The *objective* lens is the last of the three probe forming lenses.

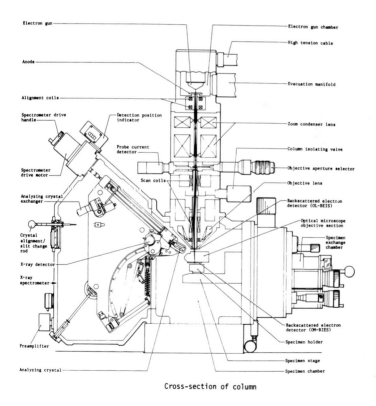

Fig. 3.1 (b) A schematic diagram of the scanning microscope interior.

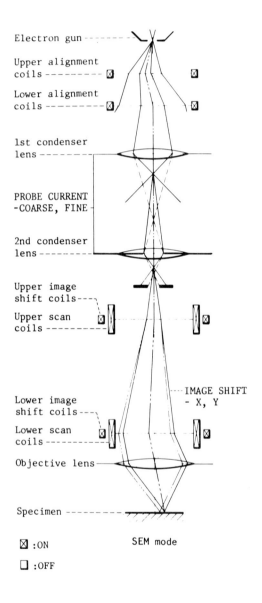

Fig. 3.1(c) A ray diagram for the conventional scanning mode ((b) and (c) reproduced by kind permission of JEOL Ltd).

I have said a certain amount in Chapter 2 about the events which befall an electron after it enters a solid specimen: I shall say a good deal more in the next chapter. Suffice it to say here by way of summary that when a solid specimen is irradiated by an electron beam the electrons which emerge fall into two energy types. The high energy ones mainly originate as beam electrons which have been elastically scattered (see section 2.9) virtually back on their tracks. They are called *backscattered* electrons. The low energy electrons (< 50 eV) originate mainly in the specimen. They are orbital electrons which have been ejected by a fast electron in the beam (see section 2.3). These electrons are called *secondary* electrons. For easily guessable reasons the secondary electrons provide information about the local orientation and position of the specimen surface (surface topography) whereas the backscattered electrons provide information about the local average atomic number of the specimen. The reason for introducing this distinction at this stage is that the two types of signal are detected differently. Typically the secondary electron detector is a scintillator which electrons cause to emit light. This passes along a light pipe and is reconverted to electrons by a photomultiplier. The typical backscattered electron detector is a silicon diode placed on the bottom of the objective lens. The secondary electron detector does not appear in Fig. 3.1(b) because it is outside of the section drawn. An optical microscope with which to view the specimen *in-situ* is often supplied. The additional backscattered electron detector is on the bottom of the optical microscope, because this obscures the normal detector when inserted into the column. Fig. 3.1(b) includes one of the total of three possible WDX spectrometers. The other spectrometer (an EDX) is not shown on the schematic but does appear on the photograph in (a) (the large container appearing at the top right of the microscope provides the cooling). Both WDX and EDX spectrometers will be described later in this chapter.

A 'pure' SEM will have no optical microscope and a short objective lens - specimen distance to improve image resolution. A pure EPMA will have a large objective lens - specimen distance, reducing the resolution but allowing for up to five WDX spectrometers and a sophisticated automatic specimen stage. There will also be a high quality optical microscope for defining specimen height for WDX spectrometry and to supplement the poor quality electron image, as well as facilities for beam current measurement and regulation.

3.2 Transmission electron microscope

Fig. 3.2 shows a photograph of a typical TEM, again with a corresponding schematic of the interior and a ray diagram for when the microscope is being operated in one of the probe modes. Several features are similar to the SEM already shown in Fig. 3.1, but in the TEM the specimen comes half-way down the column (for obvious reasons). There is the same gun (tungsten, LaB_6 or FEG as before) and condenser lenses, but the objective lens in a TEM is an image forming lens on or within which the specimen sits. The objective is the most important lens in a TEM and controls its resolution. The normal mode of operation is fixed beam, where the whole image is formed at once. The objective and subsequent lenses produce a magnified image of the specimen (or alternatively a diffraction pattern, if required). The number of lenses after the objective lens used to be two, but this number has tended to increase as microscopes have become more sophisticated.

Most TEMs will operate in scanning mode if required. In this case the upper (electron beam entrance) part of the objective lens field helps the condenser lenses to form the electron probe. Instruments designed specifically to operate in scanning transmission mode will be described in the next section.

Because in a TEM the electrons must penetrate through the specimen, TEMs are usually operated at higher beam voltages than SEMs: 100 - 400 kV as opposed to 5 - 50 kV. (They are also dearer in roughly the same proportion!) Biological specimens have very low atomic numbers (with good transmission of electrons) and are often viewed at lower beam voltages to improve contrast.

An EDX detector is attached to the microscope shown in Fig. 3.2 (a) half-way down the column, slightly to the rear. It can be seen more clearly in the inset. For a TEM, energy dispersive rather than wavelength dispersive detection is always employed. Thin specimens do not produce enough X-rays to make WDX detection practicable and in any case the very restricted space in the objective lens region makes it difficult to incorporate WDX.* The detector is anywhere between horizontal (strictly, perpendicular to the beam) and at 65° to horizontal. Because bremsstrahlung is peaked in the direction of the electron beam (see section 2.8) and contributes a substantial part of the X-ray spectrum background, the ratio of character-

---

* In fact the first TEMs incorporating an X-ray spectrometer, called EMMA, were equipped with WDX spectrometers.

istic X-ray peaks to background should improve as the X-ray detector becomes antiparallel to the electron beam.

At the bottom of the microscope there may be an electron spectrometer for analysing electron energy losses. Electron spectrometers will be mentioned briefly in Chapter 5. A typical electron energy loss spectrum can be seen on the right hand computer display in Fig. 3.2(a).

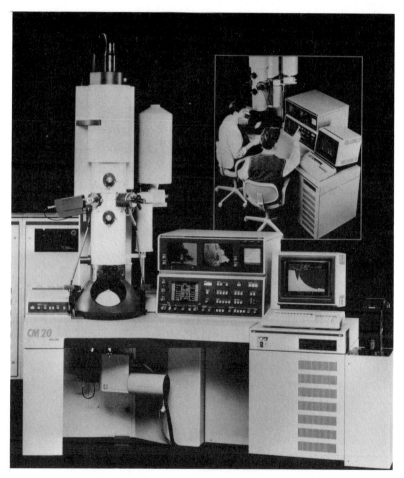

Fig. 3.2(a) A transmission electron microscope (TEM) (Philips CM20). This particular microscope has a scanning (STEM) attachment.

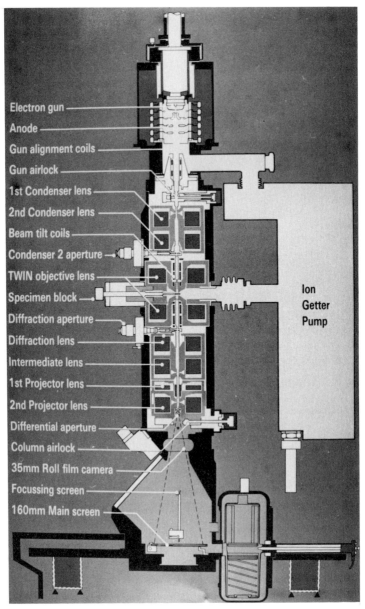

Fig. 3.2 (b) A schematic diagram of the microscope interior.

Microscopes and Spectrometers 65

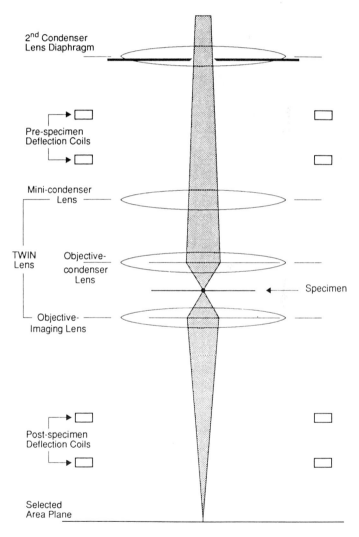

Fig. 3.2(c) A partial ray diagram for one of the probe forming modes. The image of the electron probe in the selected area plane undergoes subsequent magnification. (Reproduced by kind permission of Philips Analytical.)

3.3 Scanning transmission electron microscope

I have already said that TEMs can operate in scanning mode as STEMs. Indeed SEMs can be modified to work in transmission. Why then is it worth having a separate section devoted to STEMs? To achieve a resolution comparable with an ordinary TEM with a reasonable beam intensity, it is necessary for STEMs to be equipped with field emission guns. This imposes an exacting vacuum requirement on the microscope. Although it is possible to fit a FEG to a conventional TEM, *dedicated* STEMs - microscopes built to operate primarily in STEM rather than TEM mode - have been far more successful to date. STEMs have carved out a niche of their own at the ultra-high spatial resolution end of microanalysis.

Fig. 3.3 shows a photograph of a dedicated STEM. The principles of operation are similar to an SEM, although there are post-specimen lenses followed by the detector. The most obvious difference is the appearance of the microscope, dictated by the high vacuum requirement ($< 10^{-10}$ torr). Like TEMs, STEMs may be fitted with EDX (and electron energy loss) spectrometers.

3.4 Energy dispersive X-ray spectrometer

Fig. 3.4 illustrates the various aspects of EDX detection. EDX detectors may be seen in position in Figs. 3.1(a), 3.2(a) and 3.3.

Most EDX detectors have at their centre a silicon crystal 'drifted' with lithium,* written 'Si(Li)'. The lithium compensates for impurities, leaving effectively electrically intrinsic silicon. When an X-ray enters the crystal it is absorbed. The subsequent events are as described in Chapter 2, sections 2.6 and 2.7. The absorption process creates photoelectrons, most of them from the K shell if the X-ray energy is larger than E_K for silicon. The electronic vacancies yield X-rays and Auger electrons and the photoelectrons themselves lose energy in the ways also described in Chapter 2. The end result of this cascade of events is that a certain number of electrons is promoted from bound valency states to free conduction states, leaving behind holes in the valency states; the remainder of the energy appears as

---

*Germanium detectors are the only serious competitor. The response curve and escape peaks are rather more complicated than for silicon, but germanium detectors now have better resolution and are particularly attractive for high X-ray energies. The methods of section 3.4 can be applied equally well to germanium.

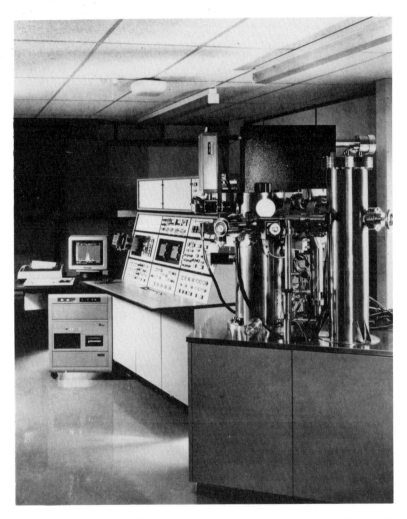

Fig. 3.3 A STEM (VG HB501). Guess which is the microscope! (In fact it is the tall cylinder on the left.) The appearance of the microscope is dictated by the vacuum necessary for a FEG source. Note the microscope is 'upside down' as compared with those in Figs. 3.1 and 3.2 - the source is at the bottom. The rectangular box on the top is an electron energy loss spectrometer. (Reproduced by kind permission of Fisons Instruments (VG Microscopes).

68 Chemical Microanalysis Using Electron Beams

heat. An electrical potential (~ 1 kV) is applied across the crystal and both the freed electrons and the holes migrate to the electrodes either side of the crystal (in opposite directions, but at roughly the same speed). The number of electrons is proportional to the energy of the X-ray, except at low X-ray energies. The little pulse of electrical current is amplified and, on the basis of its size, a given channel in a *multichannel analyser* (MCA) has its contents incremented by 1. This may be done several thousand times a second and is the basis of energy dispersive analysis of X-rays. The contents of an MCA are being displayed on a computer screen in Fig. 3.3(a).

Resolution

The peaks in the EDX spectrum of Fig. 3.4(d) correspond to characteristic X-rays. It is the sizes of these peaks which we will later turn into a chemical composition. The width of a line is usually described in terms of its *Full Width at Half Maximum* (FWHM) (see Fig. 3.5) and the term *resolution* in this context usually means FWHM. The *intrinsic* FWHM is related via the Heisenberg relationship $\Delta E \, \Delta t \sim h/2\pi$ to the time of existence of the excited state giving rise to it. This time of existence decreases as Z increases and is of the order of 10^{-16}s, leading to natural widths in cases of interest of < 4 eV. The lines in Fig. 3.4(d) are clearly much wider than this. This is for two reasons: the statistics of the X-ray absorption process in the Si(Li) crystal and electronic noise from the circuit.

The number of conduction electrons, N, created by the absorption of the X-ray photon is proportional to the energy of the X-ray, E, except at very low energies, and equals E/ε where ε is the average energy per freed bound electron and for Si(Li) = 3.8 eV. Then, for example, a Co K_α X-ray whose energy is 6.93 keV would give rise to 1819 electron-hole pairs *on average*. Because there are competing processes, leading ultimately to heat production, there is some scatter in the exact number of free electrons produced. In this case, for example, one Co K_α X-ray might give rise to 1800 free electrons and another to 1840 free electrons.

The distribution is *normal* or *Gaussian* (in effect: actually Poissonian). If the proportion of energy devoted to freeing electrons were small, the standard deviation of the distribution would be $\sqrt{N}$. If it were 100%, the standard deviation would be zero. The true answer is written as $\sqrt{(FN)}$, where F is called the *Fano factor* (the same Fano as in section 2.3); in silicon F ~ 0.1. (In fact, F is also influenced by how efficiently the freed electrons and holes are detected.) Thus the relative standard deviation is

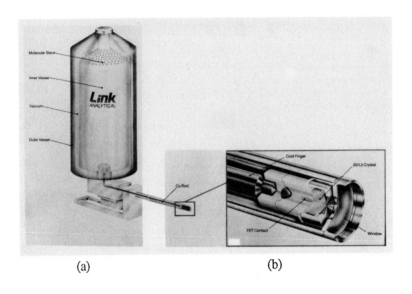

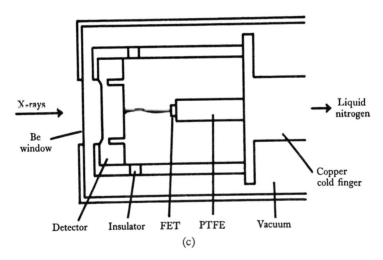

Fig. 3.4 Energy dispersive X-ray (EDX) detection. (a) Drawing of EDX detector assembly, (b) enlargement of detector and (c) schematic diagram of detector. ((a) and (b) reproduced by kind permission of Oxford Instruments. (c) is reproduced from Reed[18] by kind permission of Cambridge University Press.)

70 *Chemical Microanalysis Using Electron Beams*

σ/N = √(εF/E) and the contribution to the FWHM in eV is 2.355 σE/N

$$= 2.355 \sqrt{\varepsilon FE} \qquad (3.1)$$

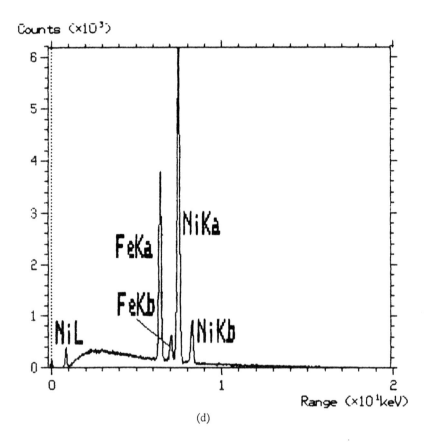

(d)

Fig. 3.4 (cont.) Energy dispersive X-ray (EDX) detection. (d) A typical EDX spectrum (see also Fig. 3.9 for WDX equivalent). A computer and display appear in Fig. 3.3.

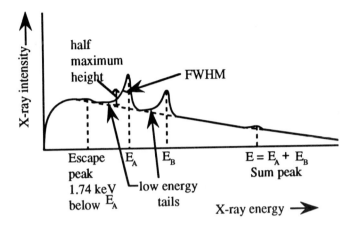

Fig. 3.5 EDX spectrum artefacts (much exaggerated). A and B are two different X-ray peaks.

The circuit noise contribution to the width of the peak is independent of energy and currently ~ 75 - 90 eV.

The three components are all mutually independent and add as follows:

$$FWHM_{total} = \sqrt{FWHM_{int}^2 + FWHM_{stat}^2 + FWHM_{elec}^2} \quad (3.2)$$

FWHM is usually quoted as that for Mn K_α (convenient because emitted by Fe^{55} radioactive sources: no need for electrons). FWHM (Mn K_α) has typically been 150 eV, but detectors with a very small $FWHM_{elec}$ are now being produced and FWHM (Mn K_α) can be as low as 130 eV.

I shall not be dealing in detail with the crucial and ingenious electronic circuitry which is used to transfer the small pulse of electrons produced by the X-ray to the multichannel analyser. The first stage, however, is a field effect transistor. To perform well this needs to be cooled. Cooling also reduces any thermal current through the detector. It is effected by a large liquid nitrogen reservoir, which is usually the most obvious aspect of an EDX detector assembly.

Dead time

It takes about 50 μs to 'deal' with each X-ray - i.e. detect it, measure its size and store it in the MCA. If another X-ray arrives before this process is complete, both X-rays must be ignored, or the answer will be some combination of the two energies. Thus the analyser is effectively switched off for this period, which is called *dead time*. This switching off is accomplished by a fast *discriminator* circuit, which does not provide sufficiently well shaped pulses for energy sorting for the MCA, but does provide advance warning of a coincidence problem of this sort. If the slow counting circuit requires (say) a time $\tau = 50$ μs to process a pulse, then, by Poisson statistics, n photons s^{-1} result in a counting rate of nexp(-nτ). Thus the fraction of dead time would be 1 - 1/e = 63% for an input count rate of $(1/50 \times 10^{-6}) = 20000$ s^{-1}.*

There is evidently no point in bombarding a Si(Li) detector with too many X-rays. There are, however, more aspects to this than mere efficiency. Discrimination becomes more difficult as the X-ray energy drops and at low energies it is advisable to increase the time constant of the discriminator circuit and to reduce the count rate. Typical upper limits for a useful input count rate are 10,000 - 20,000 counts s^{-1}.

Spectrum artefacts

Continuing with this scenario, if the two photons arrive too closely together to be distinguished by the discriminator, they will be detected as one peak whose energy is the sum of the two: hence *sum peaks* arise (see Fig. 3.5). This problem has virtually disappeared with improvements in the electronic circuitry.

Another spectrum artefact is the *escape peak* (Fig. 3.6). Most X-rays are absorbed by ionising a silicon K electron. The K X-ray that *may* result (remember the Auger electrons) may escape from the detector without itself being absorbed - particularly if it happens to be created near to the front of the detector. The measured X-ray energy is, then, the true energy minus the SiK X-ray energy (1.739 keV) and the small escape peak appears at this energy below the main peak. Obviously peaks whose energy is less than the absorption energy for silicon K (1.838 keV) will not show escape peaks. From what has already been said about X-ray absorption, the escape peak fraction will be greatest for those X-rays with energies just above the absorption energy.

More quantitatively (Reed and Ware[38]), the rate of absorption of a unit

---

*See comments at end of Example calculation 3.2.

intensity X-ray beam at a distance z below the surface is

$$\left(\frac{\mu}{\rho}\right)^i \rho z \exp\left[-\left(\frac{\mu}{\rho}\right)^i \rho z\right] dz$$

where $(\mu/\rho)^i$ is the mass absorption coefficient of silicon for the X-ray in question, of energy E, say. Further, the K edge jump ratio of silicon (= 0.94)

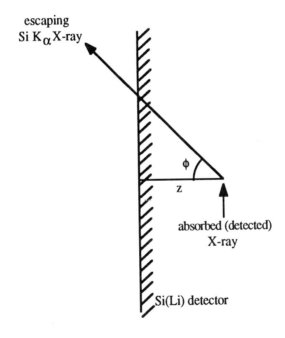

Fig. 3.6 The geometry of escape peak generation. If the detected X-ray is absorbed near the surface of the Si(Li) detector, a silicon K_α X-ray may be produced (K_β production is negligible) and may escape from the detector giving an 'escape peak' 1.74 keV below the true one.

74 *Chemical Microanalysis Using Electron Beams*

is that fraction of the ionisations belonging to the K shell. The fluorescence yield (= 0.050) defines the proportion of ionisations leading to X-rays. Finally, characteristic X-rays are emitted isotropically. Integrating the 2π steradians for which the X-rays are pointing at the surface of the detector and multiplying by a factor

$$\exp\left(-\left(\frac{\mu}{\rho}\right)^{Si} \rho \sec \phi \; z\right)$$

where $(\mu/\rho)^{Si}$ is the absorption coefficient of silicon for its own X-rays and ϕ is defined in Fig. 3.6, leads to the escape fraction ESC(E):

$$ESC(E) = \frac{r-1}{r} \omega_K \left(\frac{\mu}{\rho}\right)^i \rho z \frac{1}{4\pi} \int_z \int_\phi \exp\left(-\left(\frac{\mu}{\rho}\right)^i \rho z\right)$$

$$\times \exp\left(-\left(\frac{\mu}{\rho}\right)^{Si} \rho \sec \phi \; z\right) 2\pi \sin \phi \; d\phi \; dz$$

$$= \frac{r-1}{r} \omega_K \frac{1}{2} \left\{1 - \frac{\mu_{Si}}{\mu_i} \ln\left(1 + \frac{\mu_i}{\mu_{Si}}\right)\right\}$$

Substituting in values from Appendix A, Tables A.5 and A.6,

$$ESC(E) = 0.94 \times 0.05 \times \frac{1}{2} \left\{1 - 0.017 \; E^{2.7345} \ln\left(1 + \frac{59}{E^{2.7345}}\right)\right\}$$

$$= 0.024 \left\{1 - 0.017 \; E^{2.7345} \ln\left(1 + \frac{59}{E^{2.7345}}\right)\right\} \quad (3.3)$$

for $1.84 < E < 5.9$ keV

where E is the energy in keV of the incident X-ray. This shows that at its most significant the escape peak only really affects X-rays with energies between 1.84 and 6 keV and that even then it is small (< 1%).*

Thirdly and finally, there is a thin (~ 100 nm) region at the entrance side of the detector from where the freed electrons are not collected. This *dead layer* effectively contributes to the absorption of the X-rays before they enter the detector (of which, more below) and incomplete charge collection in general, of whatever origin, adds a low energy tail to each peak, but especially at low energies - most detector problems become worse as the X-ray energy falls (see Fig. 3.5).

Response curve of detector

Apart from escape peaks, low energy tails and sum peaks, the efficiency of the detector is less than 100% because of (a) absorption of low energy X-rays before entry to the detector and (b) transmission of high energy X-rays straight through the detector. The absorption of X-rays is by the beryllium window which separates the detector vacuum cavity from the microscope vacuum, keeping the Si(Li) crystal clean and preventing electrons and light from entering; from the gold electrode on the front of the Si(Li) and from the dead layer referred to in the previous subsection. Each of these multiplies the incident intensity by

$$\exp\left(-\left(\frac{-\mu(E)}{\rho}\right)^i \rho_i t_i\right) \quad (3.4)$$

where $(\mu(E)/\rho)^i$ is the mass absorption coefficient of the particular absorbant for X-rays of energy E. Similarly the transmission function will be

$$1 - \exp\left(-\left(\frac{\mu(E)}{\rho}\right)\rho t\right) \quad (3.5)$$

where μ, ρ and t refer to the detector.

Typical values of the thicknesses t_i are as follows:-

Beryllium window	8 μm
Gold electrode	20 nm
Silicon dead layer	150 nm
Thickness of detector	3 mm

Using these values, with the absorption coefficients of Appendix A,

---

* This estimate of ESC(E) is slightly higher than that reported by Reed and Ware[38] because the value of ω_K used here is higher.

Table A.6, results in the curve shown in Fig. 3.7.

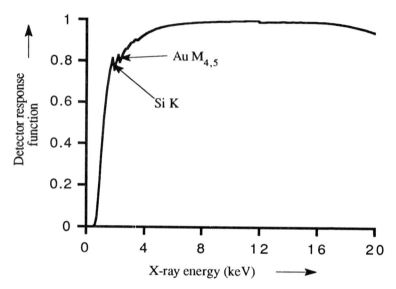

Fig. 3.7 A typical response curve for a Si(Li) EDX detector fronted by a beryllium window. Data are taken from the text and absorption factors calculated via Appendix A, Table A.6. The two most obvious absorption edges are identified.

Keep in mind that beryllium windows are in reality quite uneven and that a multitude of sins is contained within the concept of a 'dead layer'!

Thin window detectors

Low (see Fig. 3.7) and high (> 30 keV) energy X-rays are served equally badly by a Si(Li) detector with a beryllium window in front of it. The high energy peaks are no problem, because there is always an alternative X-ray (in fact L) available at lower energies. The low energy limit is more serious and prevents analysis of elements with $Z < 11$(Na). One solution to this is to replace the beryllium by a thin rigid light element compound window (e.g. BN). Just about every problem you can think of becomes worse at low

Microscopes and Spectrometers 77

X-ray energies - incomplete charge collection, non-linearity of pulse size with E, discrimination problems - not to mention absorption of the X-rays in the specimen itself. For boron and beryllium X-rays a sandwich of ~100nm of aluminium between two thin polymer sheets is sometimes used, which necessitates leaving the detector vacuum open to the microscope column.

Example calculation 3.1

The measured FWHMs for the zero energy (electronic noise) peak and the CuK_α peak on an EDX spectrometer were 88 eV and 173 eV respectively. What are the FWHMs of the Mn $K\alpha$ and Sr $K\alpha$ peaks expected to be?

From equation (3.2), ignoring the small intrinsic width contribution, for CuK_α

$$FWHM_{total}^2 = FWHM_{stat}^2 + FWHM_{elec}^2$$

$$\text{or } FWHM_{stat}^2 = \sqrt{173^2 - 88^2}$$

$$= 148.9 \text{ eV}$$

From equation (3.1)

$$\sqrt{\varepsilon F} = \frac{148.9}{2.355\sqrt{8040}}$$

where the K_α energy has been taken from Appendix A, Table A.3.

$$\sqrt{\varepsilon F} = 0.705$$

which is in reasonable accord with the numerical estimates of section 3.4 and suggests a Fano factor of ~0.13.

Then at Mn K_α (E = 5894 eV) and Sr K_α (E = 14140 eV) (Appendix A, Table A.3)

$$\text{FWHM}_{stat} = 2.355 \sqrt{\varepsilon FE}$$

$$= 128.0 \text{ eV} \quad (\text{MnK}_\alpha)$$

$$\text{and} \quad = 198.3 \text{ eV} \quad (\text{SrK}_\alpha)$$

The total widths are then

$$\text{FWHM}_{total} = \sqrt{\text{FWHM}_{stat}^2 + \text{FWHM}_{elec}^2}$$

$$= 155 \text{ eV for MnK}_\alpha$$

$$= 217 \text{ eV for SrK}_\alpha$$

Example calculation 3.2

An EDX spectrometer has a pulse processing time of 70 μs. For what input count rate is the measured count rate a maximum?

This question is concerned with the *dead time* on an EDX spectrometer. From earlier in section 3.4, $\dot{n}$ photons s^{-1} result in a count rate of $\dot{n} \times \exp(-\dot{n}\tau)$ because of successive pulses being too close to each other. The curve of measured count rate vs $\dot{n}$ will evidently pass through a maximum, whose position we require. Differentiating $\dot{n} \exp(-\dot{n}\tau)$ with respect to $\dot{n}$ and setting equal to 0:

$$\dot{n}_{max} \tau = 1 \quad \text{or} \quad \dot{n}_{max} = \frac{1}{\tau}$$

$$= 14286 \text{ counts s}^{-1} \text{ in this case}$$

There are other reasons (see section 3.1) why it may not be a good idea to operate with as high a count rate as this.

Both Example calculations 3.1 and 3.2 are more simple than in real life. In fact, count rate affects resolution and the variation of dead time with $\dot{n}$ is more complex than Example calculation 3.2 would suggest. Your manual should describe how dead time and resolution are related to energy and count rate.

Example calculation 3.3

What fraction of an Ag $L\alpha_1$ peak is its escape peak?

Recall equation (3.3):
The energy of an Ag L_α peak is 2.984 keV (Appendix A, Table A.3).

$$\text{ESC}(E) = 0.024 \left\{1 - 0.017\ E^{2.7345} \ln\left(1 + \frac{59}{E^{2.7345}}\right)\right\}$$

$$\text{ESC}(E) = 0.024 \left\{1 - 0.017 \times 2.984^{2.7345} \ln\left(1 + \frac{59}{2.984^{2.7345}}\right)\right\}$$

$$= 1.3\ \%$$

Example calculation 3.4

Above what energy of X-ray is less than one half of the intensity absorbed by a 3 mm thick Si(Li) EDX detector? (The density of silicon is 2.33 Mg m^{-3}.)

From equation 3.5

$$1 - \exp\left(-\left(\frac{\mu(E)}{\rho}\right)\rho t\right) = 0.5 \quad \text{where } \mu, \rho \text{ and t refer to the detector}$$

80 *Chemical Microanalysis Using Electron Beams*

$$\text{or } \left(\frac{\mu(E)}{\rho}\right) = \frac{\ln 2}{\rho t}$$

Substituting from Appendix A, Table A.6 and using the given data,

$$0.91309 \times 1.8389 \left(\frac{12.3981}{E}\right)^{2.94} = \frac{\ln 2}{2330 \times 3 \times 10^{-3}}$$

or E = 32.5 keV

This is roughly the K_α energy of barium. It is worth noting that the K cross-section is pretty small in the first place at this sort of energy (before worrying about actually detecting the X-rays).

Example calculation 3.5

For a Si(Li) EDX detector fronted by a 6 µm beryllium window, 15 nm of gold and a dead layer estimated at 120 nm, what fraction of sodium K_α X-rays reaches the active part of the Si(Li)? (The densities of beryllium, gold and silicon are 1.85, 19.32 and 2.33 Mg m^{-3} respectively.)

From equation 3.4, the absorption coefficient of the combined beryllium, gold and silicon dead layers is

$$\exp\left(-\left(\frac{\mu_{Be}(E)}{\rho_{Be}}\right)\rho_{Be} t_{Be}\right) \exp\left(-\left(\frac{\mu_{Au}(E)}{\rho_{Au}}\right)\rho_{Au} t_{Au}\right) \exp\left(-\left(\frac{\mu_{Si}(E)}{\rho_{Si}}\right)\rho_{Si} t_{Si}\right)$$

where E refers to the K_α energy for sodium = 1.041 keV (Appendix A, Table A.3).

From Appendix A, Table A.6

$$\left(\frac{\mu}{\rho}\right)^{Be} = 49.7 \text{ m}^2 \text{kg}^{-1}$$

$$\left(\frac{\mu}{\rho}\right)^{Au} = 555.5 \text{ m}^2\text{kg}^{-1}$$

$$\left(\frac{\mu}{\rho}\right)^{Si} = 136.3 \text{ m}^2\text{kg}^{-1}$$

The total absorption coefficient =

$\exp(-49.7\times1.85\times10^3\times6\times10^{-6})\exp(-555.5\times19.32\times10^3\times15\times10^{-9})\exp(-136.3\times2.33\times10^3\times120\times10^{-9})$

beryllium window		gold electrode		silicon dead layer
= 0.576	x	0.851	x	0.963

= 0.47

Clearly the beryllium window is the most important absorbing component. Sodium is about the lightest element which can be detected comfortably with a conventional beryllium fronted Si(Li) detector.

Example calculation 3.6

A thin window EDX detector is fitted to a TEM. Normally, from a thin foil, the ratio of the Cu K_α and Cu L_α peaks is about 1.5. Over a period of a few days, this ratio increases to about 5. It is suspected that this is due to ice on the front of the detector. Is this reasonable?

There are no data for mass absorption coefficients for hydrogen in Appendix A, Table A.6. (There are in the original paper.) Regarding water as solid oxygen with a density of 1 Mg m^{-3} and ignoring any absorption of the Cu K_α X-rays in the ice, we are asking whether,

$$\text{if } \frac{1.5}{5} = \exp\left(-\left(\frac{\mu}{\rho}\right)10^3 x_{H_2O}\right)$$

where (μ/ρ) is the mass absorption coefficient of oxygen for Cu L_α X-rays (E = 0.928 keV), is x_{H_2O}, the thickness of the ice, believable? Substituting

$$\left(\frac{\mu}{\rho}\right) = 0.461 \left(\frac{12.3981}{0.928}\right)^{2.7345}$$

$$= 552 \text{ m}^2 \text{ kg}^{-1}$$

then

$$x_{H_2O} = \frac{-\ln\left(\frac{5}{15}\right)}{552 \times 10^3}$$

$$= 2.2 \text{ } \mu\text{m}$$

This seems believable - or at least it did to us. We now have a small heater fitted in front of our detector and regular use of this has prevented the problem from recurring.

3.5 Wavelength dispersive X-ray spectrometer

Although WDX spectrometers yield graphs of X-ray intensity vs energy, just like EDX spectrometers, the method used to resolve the X-rays is fundamentally different. WDX spectrometers use the phenomenon of Bragg diffraction, referred to in section 2.6, with which you are probably familiar. As shown in Fig. 3.8, the beam of X-rays from the solid being analysed, containing the characteristic X-ray peaks and the continuous background, is allowed to fall on to a single crystal whose surface is parallel to the chosen underlying diffracting planes.

Bragg's law states that

$$n\lambda = 2d \sin \theta \tag{3.6}$$

where n is the order of diffraction (most usually and usefully 1), λ is the X-ray wavelength, d is the plane spacing and θ the angle of reflection. The

Microscopes and Spectrometers 83

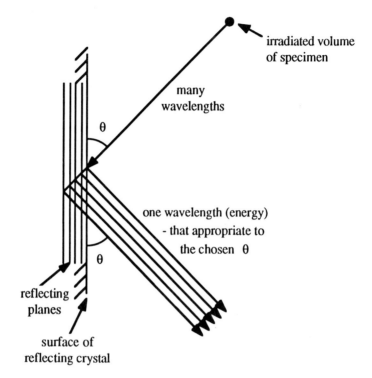

Fig. 3.8 The principle of wavelength dispersive X-ray (WDX) spectrometry. (Note the *scattering* angle is 2θ.)

crystal therefore picks out and reflects strongly one of the wavelengths, or energies, which is counted by a detector placed at the correct position.

You can see now why this is called *wavelength* dispersion rather than *energy* dispersion, although both detectors yield the same information. Notice that WDX is a *serial* technique whereas EDX is a *parallel* technique. What I mean by this is that in WDX we move from one end of the spectrum to the other, serially, whereas in EDX all energies are being counted simultaneously (not *exactly* simultaneously, obviously). Evidently parallel techniques are much more efficient - so why bother with WDX at all? The answer is that the characteristic X-ray peaks in WDX may be made much more narrow and therefore relatively much higher than those in EDX. This means that energy resolution is better and that peak to background ratios are better (roughly about 5 times those from EDX spectrometers). Fig. 3.9 compares EDX and WDX spectra from the same specimen.

Having introduced WDX in a general way, I shall now deal in a little more detail with four aspects of WDX spectrometry.

Perfect vs mosaic crystals

If a reflecting crystal of high perfection is used, a very narrow X-ray peak may be produced. If the crystal is deformed in some way, e.g. by abrasion, both the width of the peak and the reflecting power rise by an order of magnitude. Many crystals have sufficient natural internal defects not to need any extra deformation. The poorer resolution obtained from such *mosaic* crystals is perfectly adequate for the purposes of WDX spectrometry and crystals in this imperfect form are always used in practice to take advantage of the better reflectivity.

Geometry and mechanism

The simple geometry of Fig. 3.8 would be rather inefficient in practice because, with the X-ray source being effectively a point (the small irradiated volume of the specimen), the Bragg angle would be satisfied only along a curved line on the specimen surface and a strange shaped detector would be required to receive all the X-rays. What is actually done is that the crystal is bent to a cylindrical shape which 'focuses' the X-rays to a single point. The geometry is shown in Fig. 3.10. The source, crystal and detector are on the circumference of the *Rowland circle*, radius R and the crystal is bent to a cylinder of radius 2R. This is much more convenient for detection and much more efficient, too, because reflection takes place from a large area of the crystal. In fact, a toroid would be even more efficient, but too complicated technically. The fact that the source, crystal and detector have to lie in or close to a plane means that the position of the specimen is far more critical for WDX spectrometry than for EDX spectrometry.

The Rowland circle geometry actually used permits of many variations. The WDX spectrometer fitted to the microscope of Fig. 3.1 is illustrated in Fig. 3.11. It is a linear spectrometer, where the crystal travels along a

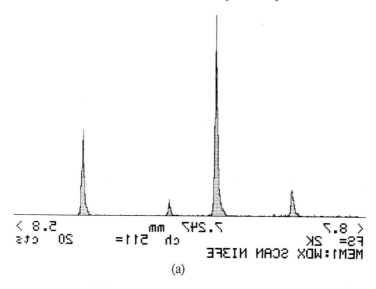

(a)

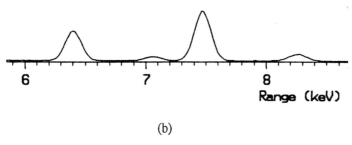

(b)

Fig. 3.9 WDX and EDX spectra from the same specimen of Ni$_3$Fe. (a) The WDX spectrum. The background is too low to be resolved clearly. (The spectrum has been printed back-to-front to facilitate comparison with the EDX spectrum.) (b) The central portion of the EDX spectrum, scaled so that the total numbers of counts in the peaks are roughly the same as for those in the WDX spectrum. The full EDX spectrum was shown earlier in Fig. 3.4(d).

86 *Chemical Microanalysis Using Electron Beams*

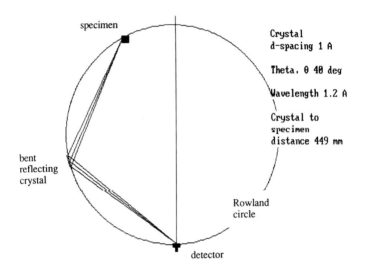

Fig. 3.10 The geometry of the Rowland circle. The reflecting crystal is bent to a cylinder of radius twice that of the Rowland circle. (Diagram based on output from Institute of Materials software series disc [reproduced by kind permission of the author, Professor F.J. Humphreys]. A = Ångstrom unit = 0.1 nm.)

straight line, rotating in a defined way as it does so. The detector follows a complicated trajectory achieved via levers. The size of the Rowland circle as implemented is a compromise between how much the crystal may be bent and how large the spectrometer is. R = 200mm is typical (here 140mm in fact).

Crystals and energy ranges

A reasonable range over which to use a given crystal is $\theta = 15° - 65°$. Above this range reflection is not sufficiently efficient and below, mechanical problems are limiting. First order reflections only are used (because they

Microscopes and Spectrometers 87

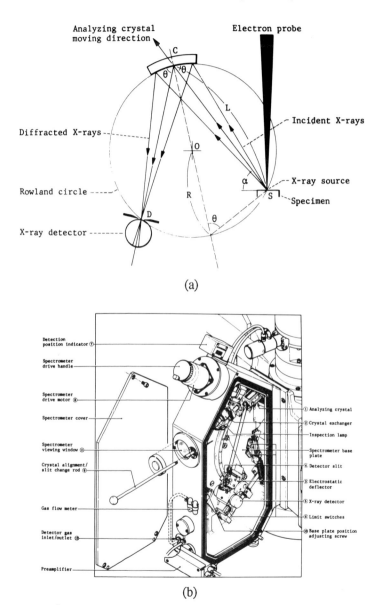

Fig. 3.11 A WDX spectrometer. (a) A schematic diagram and (b) a drawing of a real spectrometer. (By kind permission of JEOL Ltd.)

give strong reflection) and thus this defines a range of X-ray energies for each crystal. Some typical crystals and their X-ray energy ranges are shown in Table 3.1. For soft X-rays where large crystal plane spacings are required, artificially constructed crystals or ruled diffraction gratings may be used. A microscope (typically an EPMA) may have up to five spectrometers, but it will still be necessary to change crystals within each spectrometer, requiring some ingenious engineering.

Analysing crystal	Reflecting planes	d (nm)	Detectable energy range[1] (keV)
Lithium fluoride	(200)	0.2013	3.40 - 14.25
Pentaerythritol	(002)	0.4371	1.57 - 6.56
Thallium hydrogen phthalate	(100)	1.288	0.53 - 2.22
Rubidium hydrogen phthalate	(100)	1.306	0.53 - 2.20
Lead myristate	-[2]	4.00	0.17 - 0.72
Lead stearate	-[2]	5.02	0.14 - 0.57
Lead cerotate	-[2]	6.85	0.10 - 0.42

Table 3.1 Usable X-ray energy ranges for various WDX reflecting crystals.

Notes 1. This energy range corresponds to a range of θ of 12.5° - 65°
2. These are artificially constructed crystals.
(By kind permission of JEOL Ltd.)

Detectors

The detector used for WDX is a gas filled proportional counter. A schematic diagram and a photograph of a real detector appear one above the other in Fig. 3.12. The principle of operation is similar to the solid state

Si(Li) detector described in the last section, 3.4. The main differences between the two types are that the proportional counter is capable of higher counting rates (a factor of 10) but with much inferior energy resolution.

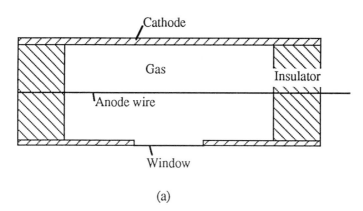

(a)

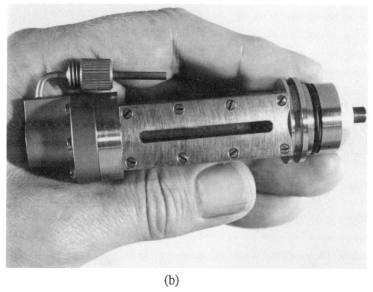

(b)

Fig. 3.12 (a) Schematic of proportional counter. (b) A real detector. The long slit is where the X-rays go in through a window. The wire is inside along the centre. The little tube coming out underneath is for gas.

90 *Chemical Microanalysis Using Electron Beams*

The only way in which the energy resolution of a proportional counter is used is to eliminate second and higher order spectra during the counting. (These would obviously have very different λ's and E's.) Just like Si(Li) detectors, gas detectors can be fitted with thin windows for low energy X-ray analysis.

Example calculation 3.7

Is PET (pentaerythritol) a suitable crystal for analysing Al K_α X-rays? (The PET crystal face is parallel to 002, for which d = 0.8742 nm.)

From Bragg's law, equation (3.6)

$$n\lambda = 2d \sin\theta$$

Since
$$E = h\nu$$
and

$$\nu = \frac{c}{\lambda}$$

then

$$\lambda = \frac{hc}{E}$$

For WDX, n is always 1. For Al K_α, E = 1.486 keV (Appendix A, Table A.3)

and
$$\lambda = \frac{6.625 \times 10^{-34} \times 2.9979 \times 10^{8}}{1.6021 \times 10^{-19} \times 1486}$$

$$= 8.3376 \times 10^{-10} \text{ m}$$

$$\therefore \quad \theta = \sin^{-1}\left(\frac{\lambda}{2d}\right)$$

$$= 28°$$

This is within the 'reasonable' range 15° - 65°.

Chapter 4   Working out the compositions

Chapter 3 described how an X-ray spectrum may be collected from a thin specimen in a TEM/STEM or from a bulk specimen in an SEM. I now wish to explain how these spectra may be converted into chemical compositions using the physics outlined in Chapter 2. There will be two main parts to this chapter: one dealing with thin film X-ray analysis and one with bulk X-ray analysis. I have chosen to deal with thin film X-ray analysis first because the correction procedures are simpler than those for bulk specimens.

4.1   Thin film X-ray analysis*

As described in Chapter 3, this is performed in a TEM/STEM or in a dedicated STEM. The TEM/STEM may be operated either in TEM (conventional, fixed beam) mode or STEM (scanning) mode. What is important is that a small stationary probe of electrons irradiates a small volume of the thin specimen. If this is done in TEM mode, the beam is gradually focused down to a small spot using the first and second condenser lens controls. If in STEM mode, the beam is merely stopped at the required position, which is usually selected by positioning an electronically generated spot, or two perpendicular crossing lines, on the CRT screen. The STEM method is more convenient for this because the image remains the same until the moment when the scanning probe is stopped on the chosen analysis position. In fixed beam, TEM mode the appearance of the spot to be analysed changes considerably as the beam is focused down, because the angle of convergence of the beam is changing (increasing). It is also usually necessary to remove the objective aperture (used for selecting a particular beam) because it will otherwise be a large source of extraneous and unwanted X-rays. This removal has the effect of reducing contrast, making it difficult to keep the chosen feature in the middle of the probe. Both the focusing down of the beam and the removal of the objective aperture may affect the position of the specimen because of the different heat input. In general, therefore, STEM operation is preferred, although it should be

---

*For good general references see Reference 39 and Williams.[40]

added that were the change from scanning to stationary beam to cause drift (although there is no change of total current input) it would not be detected.

Recall that probe sizes in a TEM/STEM equipped with a tungsten or LaB_6 gun are typically tens of nm, whereas those in a dedicated STEM will be a very few nm, and that X-ray spectrometers on transmission instruments are always EDX, not WDX. The resulting geometry is shown in Fig. 4.1. The volume of specimen irradiated by the electron beam emits characteristic X-rays (which are what we are interested in) isotropically, and bremsstrahlung rather less so, the bremsstrahlung being peaked forward in the direction of travel of the beam electrons. Thus X-ray detectors in transmission instruments are either at 90° to the beam, or look at the electron entrance side of the specimen, as shown (see also Figure 3.2).

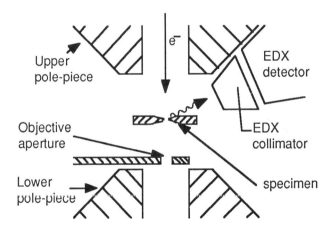

Fig. 4.1 The geometry of TEM/EDX microanalysis (schematic: adapted from the geometry for the Philips CM20 microscope illustrated in Fig. 3.2, by kind permission of Philips Analytical.)

The solid angle subtended by the EDX detector at the 'active'* part of the specimen - that irradiated by the electrons - is obviously of great importance, particularly at small probe sizes where the beam current is also small. Typical values range from 0.05 steradians to 0.15 steradians.**

Fig. 4.2 shows a typical EDX spectrum collected from $Nd_2Fe_{14}B$.

---

* *Not* in the sense of *radio* active!
** A sphere subtends a solid angle of 4π steradians at its centre.

94 *Chemical Microanalysis Using Electron Beams*

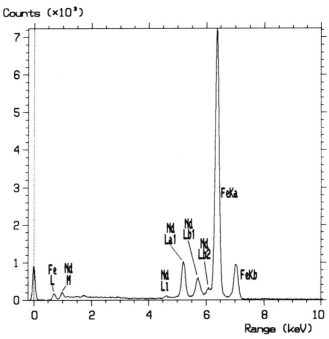

Fig. 4.2 An EDX spectrum from $Nd_2Fe_{14}B$. The main peaks showing are FeK_α and FeK_β and the Nd L series. Also visible are Nd $M_{\alpha,\beta}$ and FeL. B K at 185 eV is too strongly absorbed in the beryllium window to show.

Identifying which elements are present is fairly straightforward. The positions and relative heights of the various peaks should be stored in the microcomputer incorporated in your EDX equipment, but in any case the positions could be looked up in Appendix A, Table A.3. In Fig. 4.2 the L lines of neodymium and the K lines of iron clearly match the collected spectrum. The iron L_α and neodymium M_α lines are also visible, but the boron K lines are absorbed by the beryllium window and do not figure. An essential part of EDX microanalysis is to check the internal consistency of all the lines in the spectrum - for example, if the K lines for an element are present, are the L? etc. A good scheme is to work systematically from the high energy end of the spectrum through to the low energy end, identifying all the peaks on the way and making sure that all the peaks which should be present *are* present.

I have already mentioned that the objective aperture must be withdrawn, otherwise it will act as a source. Extraneous electrons and X-rays strike that part of the specimen not directly irradiated by the beam, as well as all the surroundings of the specimen, and excite an unwanted contribution to the spectrum. This may largely be circumvented by displacing the beam off the edge of the specimen and collecting a *hole spectrum* under the same conditions, for the same counting time. This may then be subtracted from the original spectrum before processing it. This should only be necessary when the specimen is very thin. It will not remove problems caused by high angle scattered electrons from the specimen going on to strike the lower pole piece (see Fig. 4.1) and either excite potentially fluorescing X-rays or be backscattered up towards the specimen (Li and Loretto[41]). These are best reduced by coating the surroundings of the specimen, particularly the lower pole piece, with a low Z absorbing layer.

We now wish to extract from the spectrum the chemical composition of the volume of material giving rise to it. This is what this whole section is chiefly about. The transformation of the spectrum into a composition involves two main stages:

1. The extraction of the true intensities of the characteristic X-rays detected by the Si(Li) crystal.*
2. The conversion of these intensities into a chemical composition.

I will deal with each stage in turn:

4.1.1. Spectrum → X-ray intensities detected by Si(Li) crystal

There are three steps involved in this stage, which are sometimes performed discretely and sometimes in pairs or altogether:

(i) Background subtraction

(ii) Escape peak correction

---

* 'Si(Li)' should be understood as 'Si(Li) or Ge' throughout.

(iii) Extracting the characteristic line intensities.

The *true* spectrum, as counted by the EDX detector - i.e. including the effects of window absorption and of X-rays passing undetected through the Si(Li) crystal - looks rather as in Fig. 4.3(a). The characteristic peaks have effectively zero width (see section 3.4) and the background passes through

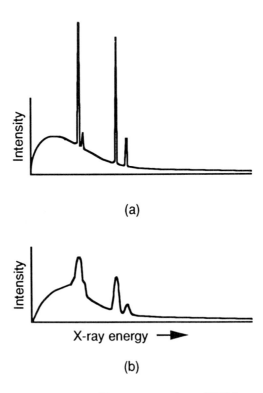

Fig. 4.3 The degradation of the X-ray spectrum by an EDX detector.
(a) The spectrum of X-rays leaving the sample. The characteristic X-rays have their natural, very small energy spread. The bremsstrahlung background in principle rises monotonically as its energy decreases, but specimen self-absorption (more marked in bulk specimens in fact) creates the shape shown.
(b) The spectrum as detected by EDX. Note the energy spread of each characteristic peak, the low energy tails and the added absorption at low energies caused by the protective beryllium window.

a smooth maximum (created by absorption of the long wavelength components in the window and the specimen itself) on which may be superimposed small discontinuities corresponding to absorption edges for the gold electrode and silicon 'dead layer'. If the spectrum really looked like this (and a WDX spectrum does, almost) then it would be a simple matter to read off the numbers of counts, or detected X-rays, corresponding to each characteristic X-ray energy, make a small escape peak correction, and the purpose of section 4.1.1 would be accomplished. Unfortunately, EDX spectra actually look like the example shown in Fig. 4.3(b), because each peak has a (very) finite width and a non-Gaussian tail (see section 3.4).

When an instrument degrades a signal, the usual way of *unfolding* or *deconvolving* the original signal from the final signal is to Fourier transform both the measured signal and the instrumental response function, divide the first by the second and then transform back again. In principle this should produce a spectrum like that in Fig. 4.3(a) from one like that in Fig. 4.3(b). Accepting that the effect of the instrument varies across the spectrum so that we have to be a little more clever than just described, this approach is the basis of much successful spectrum processing. In fact the spectrum is generally multiplied electronically by a 'top-hat' function as shown in Fig. 4.4. This will obviously remove background and in fact produces a differentiated spectrum. The detector response is obtained by using a series of specimens each of which contains an *isolated* peak (ideally from an element): this will define the EDX detector response at a series of energies. (This is very like obtaining a Green's function at successive positions across the spectrum.) These isolated peaks are similarly transformed. The transformed peaks are then *matched* to the transform of the spectrum to be analysed (equivalent to *dividing* two transforms). This procedure takes care of background and (in principle) escape peaks as well as peak integration.

The reasons why it is not automatically the answer to every EDX problem are:

(i) The shape of the background depends on the matrix composition in what may be a complicated way. The shape of the background immediate to the peak - for example under the low energy tail - may be different for the element from what it is for the compound or alloy.

(ii) Neighbouring peaks overlap. The transforms tend to be nar-

98 Chemical Microanalysis Using Electron Beams

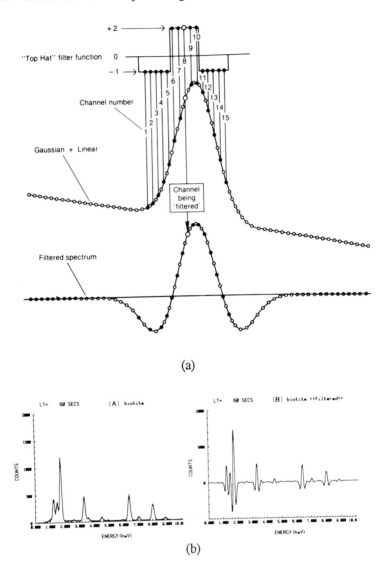

Fig. 4.4   Spectrum processing by digital filtering.
(a) The spectrum is multiplied by a 'top-hat' filter function.
(b) An example of a spectrum and its transform.
(Taken from Williams[40] by kind permission of TechBooks.)

rower, but may still need disentangling from each other. This may be performed by least squares fitting of a linear combination of the two or more spectrum transforms contributing to the final spectrum.

(iii) All of these operations, including (ii) above, are straightforward, in principle, on perfect spectra. Of course, real spectra display channel-to-channel statistical variations (see Fig. 4.4). It is not obvious how the procedure described will cope with imperfectly defined data. Not surprisingly, problems sometimes arise.

A different approach from that outlined above, which does not use transforms, involves performing the three steps separately:

You may recall (section 3.4) that escape peaks are a result of Si K photons excited in the Si(Li) crystal escaping from the detector without being themselves detected. The appropriate formula is equation (3.3). To be exact this should be applied to a spectrum in its undegraded (Fig. 4.3(a)) form. The escape peak correction, however, is very small anyway, and little inaccuracy is introduced by applying it channel by channel. It is convenient, therefore, to perform it right at the outset. Like most of the operations described in this first spectrum processing section, it is best done using the computer and proprietary software which should be supplied with your EDX detector. It would certainly be very tedious to do by hand on, for example, an MCA with 2000 channels.

The background may be modelled using a mathematical expression. If the exact version of equation (2.15) (i.e. including the constant of proportionality and the quantum mechanical and directional correction terms (Bethe and Heitler[29] theory must be used at TEM beam voltages)) is multiplied by the absorption and transmission functions of the detector, equation (3.5), then the resulting expression underestimates the experimentally observed background levels, particularly at low energies. This is because the background not only consists of bremsstrahlung from the specimen but also of bremsstrahlung from the surroundings of the specimen within the microscope as well as the effect of high energy electrons which penetrate the beryllium window. The bremsstrahlung from the microscope interior is accompanied by highly undesirable characteristic peaks which will clearly spoil any attempted analysis. These X-rays are much reduced by the use of appropriately positioned thick (or 'top-hat') apertures in the condenser system of the microscope and of collimators on the end of the

EDX detector. The penetrating high energy electrons are a grave problem because they degrade the detector, which has to be warmed up to anneal out the damage. This is a nuisance and can also lead to lithium diffusion and permanent detector malfunction. For this reason manufacturers now allow EDX detectors to be retracted, or protected by a shutter, when this sort of problem threatens - especially on 'medium voltage' 300-400 kV instruments. For more information on how to identify and eliminate these sorts of problem, see Williams[40] and Zaluzec.[42] Even when the 'microscope spectrum' has been reduced as much as possible it is good practice to record and subtract from the specimen spectrum a 'hole spectrum', collected with the beam off the edge of the specimen.

Returning to the modelling of the background, although the true background really contains more than just specimen bremsstrahlung, the Kramers[27] expression, equation (2.15), multiplied by the detector absorption / transmission function is scaled to the background at positions away from characteristic peaks. In fact the Kramers expression for a *bulk* specimen, described below, is sometimes mistakenly used but the energy dependence is quite similar to that of equation (2.15) for a high beam voltage, such as is found in a TEM, and little harm is done thereby. Such a match is shown in Fig. 4.5. Various terms can be added to the basic Kramers expression to improve the fit. Just as there were problems with the filtering method, so the perfect match does not exist and there is always some uncertainty - particularly in the important low energy region.

Finally, the peaks may be matched, using some least squares method, either to stored spectra or to computed profiles which should include the Gaussian central part of the peak and the non-Gaussian low energy tail. Alternatively, electronic *windows* can be used to select those MCA channels corresponding to a fixed proportion of the peak. The usual criterion for window width is to use 1.2 FWHM (= Full Width at Half Maximum). Clearly where there is peak overlap a more complicated least squares fit is required, similar to that adverted to above in the discussion of peak transform overlap.

I have described two alternative systems for spectrum processing. Although I have treated them in isolation, parts of each system may be mixed - for example the escape peak may be removed before transforming the spectrum (in which case the escape peaks would not be collected with the standard single element spectra) etc.

In fact in this part of the process of deriving a chemical composition from your spectrum you are very much at the mercy of whatever system you

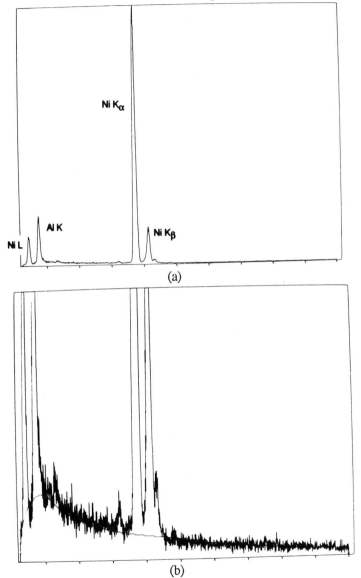

Fig. 4.5 Matching the spectrum background.
(a) The original spectrum (from Ni_3Al).
(b) A close-up of the matched background. The x axis is X-ray energy and the y axis numbers of counts.

happen to be using. You should be aware, however, exactly what your system actually does and any approximations which may be inherent. If you cannot find the answer in the manual, don't be afraid to ask the manufacturer. He will probably be quite pleased you are taking an interest and will help you get the best out of his no doubt excellent equipment.

Now for the second part of the process of deducing a chemical composition:

4.1.2 X-ray intensities → chemical composition

At this stage we imagine that intensities corresponding to a series of X-ray lines have been extracted from the measured spectrum. We now require to turn these into a chemical composition for the irradiated volume. This is easier in a very thin, than in a thick, specimen because in the very thin specimen the processes of X-ray absorption and X-ray fluorescence by the specimen may safely be ignored. I will therefore begin with very thin specimens. What exactly is meant by 'very thin' will transpire in the subsequent sections on thicker specimens.

4.1.2.1 Very thin specimens

There are only two corrections to make here. These relate to the cross-sections for X-ray excitation, and absorption in, or transmission through, the detector window + gold electrode + dead and active Si layers. Both of these factors mean that the measured X-ray intensities are not in the same ratio to each other as the atomic concentrations of the elements giving rise to them. Because of the thinness of the specimen, we can neglect within reasonable accuracy:

   1. Energy losses by the beam electrons.

   2. Scattering of the beam electrons.

   3. Absorption of the characteristic X-rays on their way through the specimen towards the detector.

4. Fluorescence of one characteristic X-ray wavelength by another.

5. Fluorescence of a characteristic X-ray wavelength by continuum (background) X-rays.

For thicker specimens (section 4.1.2.2.) we have to include 3 and 4 in the list above (and for the purposes of spatial resolution only, 2) but not 1 and 5. For bulk specimen analysis in an SEM all of 1 to 5 have to be considered.

There are two ways of making the two necessary corrections:

Method 1 : Using a standard

The safest and most satisfactory way of correcting for cross-section and window absorption / transmission effects is to use as a standard, or calibrator, a compound of known composition, which contains the elements of interest.

Example calculation 4.1

An EDX spectrum from a thin specimen containing nickel and aluminium yields the following counts after subtraction of a hole spectrum and background removal:

Ni K_α	Al K	
9736	5355	Ni/Al = 1.82

A spectrum from a thin specimen of Ni_3Al was found to contain

Ni K_α	Al K	
11668	2030	Ni/Al = 5.75

What is the unknown composition?

Since for Ni_3Al,
$$\frac{Ni}{Al} = 3$$

the unknown composition is defined by

$$\left(\frac{Ni}{Al}\right)_{unknown} = 3 \times \frac{1.82}{5.75} = 0.95 \quad \text{i.e. Ni - 51.3 at\% Al}$$

(This answer is WRONG!)

Those of you who know something about the nickel - aluminium phase diagram must at this moment be getting ready to write to the publishers of this book, or to the author, to point out that Ni_3Al can have a whole range of compositions. Indeed it can, and I deliberately* chose this example to demonstrate one of the many pitfalls connected with choosing a good standard. If you use an intermetallic, use a 'line' compound (one with a small homogeneity range) and even then be suspicious of it. Is it single phase? Is the phase diagram correct? In fact, in the example above, the standard had a composition Ni - 23.6 ± 0.1 at % Al as determined by wet analysis. The correct answer, then, was Ni - 49.7 at % Al rather than the 51.3% given above. (The third significant figure is optimistic in the extreme.) Incorrect use of standards can lead to much grosser errors than the one illustrated here. If desired, atomic % may be converted to weight % using the relative atomic masses of Appendix A, Table A.2. Thus Ni - 49.7 at % Al corresponds to Ni - 49.7x26.98/(50.3x58.69+49.7x26.98)
= Ni - 31.2 wt % Al.

Note that in the example given, the whole of the K peak was used for aluminium, but only the K_α for nickel. This was an arbitrary decision - K_β could have been included easily enough. It is important, though, to do exactly the same for both the unknown specimen and the standard. Intermetallic compounds are just one possible example of a standard. Other common ones are inorganic compounds - for example minerals - and artificial glasses made to a specific composition.

The whole essence of this method is in the standard. Once this has been prepared satisfactorily (see p. 119: polishing films) everything is fairly straightforward. Designing and preparing a good standard is, however,

---

*None of my colleagues believes this!

often difficult and time consuming.

Method 2 : Calculation from first principles

Sometimes a suitable standard is impossible to find, or at least very difficult. Sometimes one needs the answer to a few % only. In these circumstances the composition may be deduced directly via a calculation.

The two corrections involved are for X-ray excitation cross-sections and for detector response. To calculate the excitation cross-sections we need the ionisation cross-section, equation (2.4) or (2.5), the fluorescent yield, equation (2.11), and, perhaps, the partition factor (Appendix A, Table A.5). For the detector term we need equations (3.4) and (3.5).

Example calculation 4.2

For the data recorded in Example calculation 4.1, calculate the composition of the specimen. The analysis was performed using a TEM operating at 100 kV.

From Example calculation 2.3, the ratio of the Ni to Al K ionisation cross-sections for a beam voltage of 100 kV is 1/9.7.

From Appendix A, Table A.4, the fluorescent yields are

$$\omega_K(Al) = 0.039$$
and $$\omega_K(Ni) = 0.406$$

From Appendix A, Table A.5 the partition factor

$$\frac{K_\alpha}{K_\alpha + K_\beta} \text{ for Ni } (Z = 28) = 0.896 - 6.575 \times 10^{-4} \times 28$$

$$= 0.878$$

Using for the moment the detector response function of Fig. 3.7, itself based on the data quoted below equation 3.5, 0.698 of the AlK X-ray intensity is absorbed and 0.986 of the NiK$_\alpha$. Thus the ratio of X-ray intensities emitted per atom per electron is:

$$\frac{\text{Ni K}_\alpha}{\text{Al K}} = \frac{0.406 \times 0.878 \times 0.986}{9.7 \times 0.039 \times 0.698}$$

$$= 1.33$$

Recall from Example calculation 4.1 that the measured Ni/Al X-ray count ratio was 1.82. This implies a chemical composition of at % Ni / at % Al = 1.82/1.33 = 1.37, or Ni - 42 at % Al. This is quite different from the value of Ni - 50 at % Al deduced using a standard. This sort of error is not untypical of situations where reasonable differences in Z exist between the elements present and no special effort has been made to optimise the window parameters.

Had Method 2 of the two listed above been the only one to be described it would have made more sense to have included the absorptive and transmissive effects of the detector in with the other effects of the detector on the signal: peak width, escape peaks, etc. The reason I have put detector absorption and transmission in with the cross-section correction is because they are all necessarily combined in the 'standards' approach and it seemed clearer to reproduce exactly the same parts of the correction procedure in parallel correction calculations.

Of course, once the ratio of efficiencies of X-ray production for nickel and aluminium has been measured, it is a microscope constant and may be used thenceforward for analyses involving nickel and aluminium (assuming the detector does not change). Extending this approach, a curve of X-ray efficiency factors may be produced for all elements, typically ratioed to silicon or, more sensibly, iron (less absorption in the beryllium window and specimen). Many examples of such curves, usually labelled as k factors,[43] have been published - see for example Williams.[40] Note that for historical reasons connected with bulk specimen analysis they are usually expressed in ratios of wt %. Keep in mind also that specimen thickness and polishing films, for example (see section 4.1.2.2) must be avoided or allowed for.

The alternative but essentially equivalent approach is to fit detector parameters to experimental data.[13] In principle this approach could then be employed at any beam accelerating voltage. In fact, in the case cited above an unrealistically thick detector window would be predicted and clearly some of the other data are in error. (The most likely culprit is the aluminium

K fluorescence ratio.) On a related point it is worth checking from time to time the ratio of two X-ray intensities, one of reasonably low energy, from a known specimen. An even simpler check is to use the K and L X-rays from a thin copper specimen. This monitors several aspects of detector function - in particular the cleanliness of the window. Also worth monitoring are the FWHMs (widths) of the two peaks.

For the best results, if standards are used, they must be analysed as near in time as possible to the unknown specimen and under as similar conditions as possible (beam voltage, type of stage, tilt of stage). Even when the standard is in an adjoining recess in the same stage there is still scope for some change in, for example, the number of backscattered electrons, the amount of microscope-originated fluorescence etc. Obviously one just has to do the best one can.

4.1.2.2 Thicker specimens

As the specimen thickness increases, so we have to start worrying about differential absorption of the X-rays in the specimen, and about fluorescence of one of the characteristic X-ray wavelengths by another. Of these two effects, by far the more important and common is differential absorption, and I shall deal with it first. In both cases the corrections and procedures described in the previous section (4.1.2.1) for thin specimens need to be performed as well, and I shall assume therefore the contents of section 4.1.2.1.

Absorption correction for transmission specimens

In a specimen whose thickness is not trivially small the lower energy X-rays are absorbed more by the specimen itself than are the higher energy ones. Thus, whether the cross-section and detector absorption/transmission corrections are being carried out via the 'standards' or the 'calculation' way, it is necessary to correct the relative intensities of the X-rays to account for their different degrees of absorption in the specimen.

Fig. 4.6 shows the geometry for a common situation where the detector, electron beam and foil normal are all in the same plane. This may not be true for your microscope: I shall return to more complicated geometries later on. The specimen is tilted by θ_s towards the detector which itself is at an angle

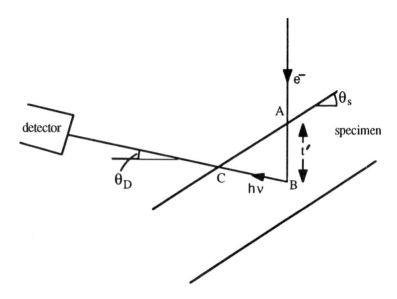

Fig. 4.6 The absorption correction for medium thickness specimens. Here the beam, specimen surface normal and detector are all in the same plane (not always true). Absorption of the X-rays occurs over a path length $BC = t' \cos\theta_s / \sin(\theta_s + \theta_D)$.

θ_D above the horizontal as drawn. θ_D is often called the *take-off angle*. At a depth t' in the foil, the distance a generated X-ray must pass through the specimen on its way to the detector will be t' times some geometrical factor which we will write as α for the time being. Since the X-rays are generated uniformly through the foil (no energy loss, no beam electron scattering) the measured X-ray intensity for element A is multiplied by

$$\frac{\int_0^t \exp(-\mu_A \alpha t') \, dt'}{\int_0^t dt'} = \frac{1}{\mu_A \alpha t} \{1 - \exp(-\mu_A \alpha t)\}$$

where I have used, for simplicity, μ_A, the absorption coefficient of the specimen for the characteristic X-rays from element A, instead of $(\mu_A/\rho)\rho$.

The variable is the foil thickness parallel to the beam. Thus for the ratio of two X-rays A and B, the ratio of the measured intensities I_B/I_A would have to be multiplied by

$$\frac{\mu_B}{\mu_A} \frac{\{1 - \exp(-\mu_A \alpha t)\}}{\{1 - \exp(-\mu_B \alpha t)\}} \qquad (4.1)$$

Note that for small μ or t the correction disappears (becomes 1) to give the ratio which would be measured from a thin specimen as common sense dictates it must. Note also that to make this correction we need both the foil thickness and the specimen density (because it is (μ/ρ), the *mass* absorption coefficient, which is tabulated). This is rather a nuisance, although like most things the speed with which you will perform this correction will improve with practice. For the geometry shown in Fig. 4.6, α is easily calculated. Noting that the true foil thickness t_{true} is the projected, or measured, foil thickness multiplied by $\cos \theta_s$, then α is evidently $\cos \theta_s / \sin (\theta_s + \theta_D)$ and the absorption correction is

$$\frac{\mu_B}{\mu_A} \frac{\left\{1 - \exp\left(-\mu_A \frac{\cos \theta_s}{\sin (\theta_s + \theta_D)} t\right)\right\}}{\left\{1 - \exp\left(-\mu_B \frac{\cos \theta_s}{\sin (\theta_s + \theta_D)} t\right)\right\}}$$

$$= \frac{\mu_B}{\mu_A} \frac{\left\{1 - \exp\left(-\mu_A \frac{t_{true}}{\sin (\theta_s + \theta_D)} t\right)\right\}}{\left\{1 - \exp\left(-\mu_B \frac{t_{true}}{\sin (\theta_s + \theta_D)} t\right)\right\}} \qquad (4.2)$$

If the plane containing the detector and the electron beam is rotated by ϕ relative to the plane containing the beam and the foil normal, then the correction is given to sufficient accuracy by multiplying $\sin (\theta_s + \theta_D)$ in equation 4.2 by $\cos \phi$. Thus the correction, in terms of t, becomes

$$\frac{\mu_B}{\mu_A} \frac{\left\{1 - \exp\left(-\mu_A \dfrac{\cos\theta_s}{\sin(\theta_s + \theta_D)\cos\phi} t\right)\right\}}{\left\{1 - \exp\left(-\mu_B \dfrac{\cos\theta_s}{\sin(\theta_s + \theta_D)\cos\phi} t\right)\right\}} \qquad (4.3)$$

Remember that μ will usually be written $(\mu/\rho)\rho$ and so in order to make a correction for absorption we need, as well as the thickness of the specimen (in the direction of the electron beam), the density also. The density of the specimen will depend on its composition, which is what we are trying to find. To find two values of these variables which are consistent with each other it is clearly necessary, in principle, to *iterate* the calculation (i.e. repeat it until neither value has to be changed). I say 'in principle' because even if, to be optimistic, the density of the sample is known as a function of composition, the correction is normally so small that the density corresponding to the composition deduced without using an absorption correction is sufficiently accurate.

For a given position on your specimen, the more you tilt the surface of the specimen towards the detector, the smaller the absorption correction will be (to be more exact, the closer to 1 will be expression (4.2), for example). This is true even though the thickness of the foil will increase as $t_{true} / \cos\theta_s$. You may like to show this to your own satisfaction from expression (4.2) by considering the way the absorption correction varies between $\theta_s = 0$ and $90°-\theta_D$ (both of which limits may be impracticable for small θ_D). In fact there may be geometrical line-of-sight reasons why you may not wish to tilt your specimen too much.

This leaves the specimen thickness to be measured. The specimen thickness is needed not only for the absorption correction but also for the fluorescence correction and an estimate of electron scattering (or *beam broadening*). These topics are dealt with later, and the following section is therefore useful for more purposes than the immediate context.

Measurement of specimen thickness in a TEM/STEM

There are two relatively common methods for measuring specimen thicknesses in a TEM or STEM. One of them involves EELS (Electron Energy

Working out the Compositions

Loss Spectroscopy) and is described in Chapter 5. The EELS method works for both crystalline and non-crystalline specimens, but obviously requires a spectrometer to be fitted to the microscope.

The second method is for crystalline specimens only and therefore involves an interaction between the electrons and the whole specimen (or at least crystal) at once (see section 2.10). A *convergent beam diffraction pattern* is taken with the crystal set at a moderately (but not very) low order Bragg position (e.g. 311 for an fcc metal). A convergent beam diffraction pattern is a diffraction pattern taken with a deliberately high beam convergence angle. The quality of the pattern is much improved if the beam is focused to a small probe, as is usually the case anyway when performing chemical microanalysis. The resulting pattern looks like that shown in Fig. 4.7. The fringes across the diffracted beam disc correspond to a periodic variation in intensity of the diffracted beam as the incident beam is tilted across the Bragg position, which is the symmetry plane across the middle of the disc. The method relies on the fact that, as shown by two beam dynamical diffraction theory (Kelly et al.[44]), for each of the minima

$$\left(s^2 + \frac{1}{\xi_g^2}\right) t^2 = n^2$$

where n is numbered sequentially from the first fringe, for which n is t/ξ_g modulo 1.* s is the *deviation parameter* at the position of the fringe in question and is defined in Fig. 4.8(a). ξ_g, the *extinction distance*, is an inverse measure of the scattering power of the **g** planes. It should be a fixed number but unfortunately is affected by other reflecting planes. Luckily we do not need to know it.

Rearranging:

$$\left(\frac{s}{n}\right)^2 = \left(\frac{1}{\xi_g}\right)^2 \left(\frac{1}{n}\right)^2 + \left(\frac{1}{t}\right)^2 \qquad (4.4)$$

A plot of $(s/n)^2$ vs $(1/n)^2$ will therefore yield a straight line whose (extrapolated) ordinate intercept is $1/t^2$.

---

*The largest integer less than t/ξ_g.

112 *Chemical Microanalysis Using Electron Beams*

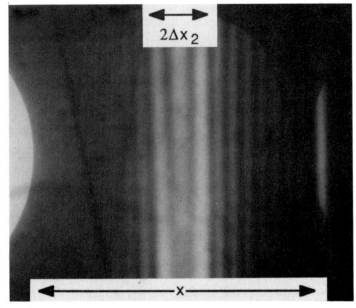

Fig. 4.7 A convergent beam electron diffraction pattern for determining specimen thickness. The disc in the centre is the dark field disc. The beginning of the main beam disc can be seen to the left, allowing x ($\equiv$ g - see text) to be measured. x is the distance between equivalent points on two neighbouring discs, not the diameter of a disc. Note that the fringes in the diffracted beam disc are symmetrical about the central line, corresponding to the Bragg orientation. The thickness clearly varied across the electron beam probe. The measurements used in Example calculation 4.3 were taken across the equator of the disc. (The specimen was Zr_3Al and the reflection a {311}.)

Example calculation 4.3

Fig 4.7 shows a convergent beam diffraction pattern consisting of a 311 reflection of stoichiometric Zr_3Al. What is the foil thickness? (The beam voltage was 400 kV. The lattice parameter of stoichiometric Zr_3Al is 0.4372 nm.)

From the symmetry and appearance of the diffracted beam disc, the Bragg position evidently corresponds to the bright centre line. From simple

Working out the Compositions 113

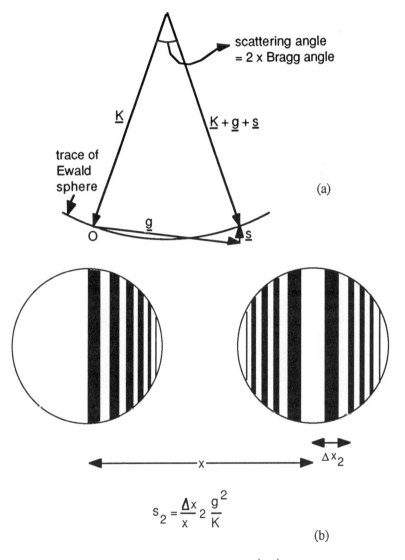

Fig. 4.8 (a) Definition of the deviation parameter s. $|K| = \lambda^1$ (see also Fig. 2.13 and Appendix A for λ formula). s as shown here is negative (positive where **g** is inside the Ewald sphere). **g** is a special case (for crystals) of **q** as used in Fig. 2.13.

(b) Working out 's' for the minima in the diffracted beam disc of a convergent beam pattern. Δx as exemplified here is for the second minimum.

114 Chemical Microanalysis Using Electron Beams

geometry (Fig. 4.8(a) and (b)) $s = \Delta x \, g^2/(xK)$ where x is the distance on the photograph corresponding to **g** and Δx is the distance between the Bragg position and the minimum in question. Table 4.1 lists the values of s for the

2Δx (mm)	s (nm⁻¹)	n^{-2}	$(s/n)^2$
5.0	0.0038	0.250	3.52x10⁻⁶
14.0	0.0105	0.111	1.23x10⁻⁶
21.0	0.0158	0.062	1.55x10⁻⁶
27.4	0.0206	0.040	1.69x10⁻⁵
34.0	0.0255	0.028	1.81x10⁻⁵
40.0	0.0300	0.020	1.84x10⁻⁵
46.4	0.0348	0.016	1.90x10⁻⁵
52.4	0.0394	0.012	1.91x10⁻⁵

Table 4.1 Values of s measured from Fig. 4.7 and used in Fig. 4.9 to deduce the foil thickness. n^{-2} and $(s/n)^2$ for the correct starting values of n (=2) only are shown.

measured minima. Fig. 4.9(a) shows various plots of the resulting points corresponding to different guessed/estimated values of n. Notice that the line is curved one way for n too small and the other for n too large. Graphs for n = 2, 3 and 4 are shown magnified in Fig. 4.9(b). The best straight line is for a starting value of n = 2. Fig. 4.9(c) shows a linear regression on all points except that corresponding to the first minimum, which tends to be less trustworthy in this respect. From the ordinate intercept (= $1/t^2$) the foil thickness is 224 nm. In this case I chose a position next to a coherent twin boundary and the projected width of this enabled me to confirm this value of foil thickness to within the accuracy with which I was able to measure. Note that here I knew what **g** and its size were, but this is not necessary if the camera length is known. Again, here I have used a photograph. The method can be applied online, however, if the microscope is fitted with a suitable measuring device - for example calibrated beam shifts. This is a very useful thing to be able to do. The convergent beam method of determining foil thicknesses has been thoroughly tried and tested. It is easy

Working out the Compositions 115

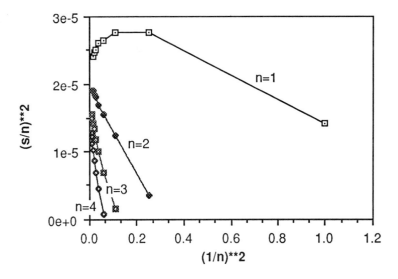

Fig. 4.9 (a) Plots of $(s/n)^2$ vs $(1/n)^2$ for various choices of n corresponding to the convergent beam pattern of Fig. 4.7. n as labelled on the graph refers to the first dark fringes. The ordinate intercept = $1/t^2$ and the gradient = $-1/\xi_g^2$

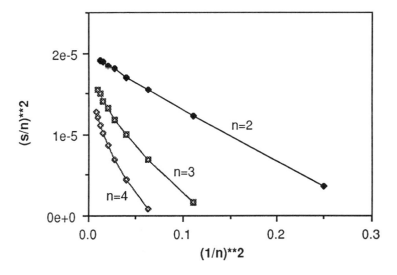

Fig. 4.9 (b) A magnified version of (a) with starting values of n = 2, 3 and 4.

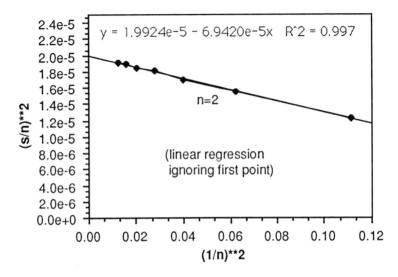

Fig. 4.9 (c) Linear regression for a starting value of n = 2. The first point (on the right) has been excluded.

and robust and is much to be recommended.

Neither of the two methods works very well in very thick foils. If you are working with a foil thickness too great for these methods, one solution is to use it on a thinner part of the specimen (if it exists) and then use the relative heights of the high energy bremsstrahlung for a given counting time to obtain the greater thickness by ratio.

Other more specialist methods for determining foil thickness exist under certain circumstances - for example the projected width of a stacking fault on a known plane. One method I suggest you *don't* use is the separation of two contamination marks caused by the beam either side of the specimen. For one thing it gives the wrong answer (always overestimates) and in any case if you are getting so much contamination that you can 'measure' foil thickness with it, you need a new microscope.

Example calculaton 4.4

A spectrum was collected from the position on the Zr_3Al specimen where the thickness was measured to be 224 nm in Example calculation 4.3. After subtraction of a hole spectrum and background removal the ratio of ZrK_α to AlK counts was 2.7. What would the ratio have been in a very thin foil? (The spectrum was collected in a JEOL 4000 FX TEM operated at 400 kV with the specimen horizontal. The detector used (there are two on this microscope) was at 65° above the horizontal.)

The spectrum must be corrected for the effects of self-absorption of the X-rays by the specimen itself. Primarily this is absorption of the soft aluminium K X-rays (energy 1.486 keV) by the medium atomic number zirconium atoms. From Table A.6, Appendix A, the full table of the mass absorption coefficients involved is

Absorbed X-ray	Absorbing atom	
	Al	Zr
Al K	38.8	173.2
Zr K_α	0.7	2.1

(all in m² kg⁻¹)

From Table A.2, Appendix A, the elemental mass concentrations are

$$C_{Zr}^m = \frac{3 \times 91.22}{3 \times 91.22 + 26.98} = 0.9103$$

and $C_{Al}^m = 0.0897$

Using equation (2.14),

$$\left(\frac{\mu}{\rho}\right)_{AlK}^{spec} = 0.9103 \times 173.2 + 0.0897 \times 38.8$$

$$= 161.1 \ m^2 \ kg^{-1}$$

and $\left(\dfrac{\mu}{\rho}\right)^{spec}_{ZrK_\alpha} = 0.9103 \times 2.1 + 0.0897 \times 0.7$

$$= 2.0 \ m^2 \ kg^{-1}$$

Using the lattice parameter of Zr_3Al given in Example calculation 4.3, its density = 5.974 Mg m^{-3}.
Then

$$\mu_{AlK} = 9.624 \times 10^5 \ m^{-1}$$

and $\quad \mu_{ZrK_\alpha} = 1.19 \times 10^4 \ m^{-1}$

Referring to equation (4.2),

$$\alpha = \dfrac{1}{\sin \theta_D} = \dfrac{1}{\sin 65°} = 1.173$$

and $\quad \mu_{AlK} \alpha t = 9.624 \times 10^5 \times 1.173 \times 224 \times 10^{-9} = 0.2529$

$$\mu_{ZrK_\alpha} \alpha t = 1.19 \times 10^4 \times 1.173 \times 224 \times 10^{-9} = 3.127 \times 10^{-3}$$

(See equation (4.1): α is used in two unconnected ways in these equations.)
Then the effect of specimen self-absorption is to increase the ZrK_α : AlK ratio by

$$\dfrac{\mu_{ZrK_\alpha}}{\mu_{AlK}} \cdot \dfrac{\{1 - \exp(-\mu_{AlK} \ \alpha t)\}}{\{1 - \exp(-\mu_{ZrK_\alpha} \ \alpha t)\}} = \dfrac{1.19 \times 10^4 \{1 - \exp(-0.2529)\}}{9.624 \times 10^5 \{1 - \exp(-3.127 \times 10^{-3})\}}$$

$$= 0.88$$

The ratio in a thin foil would therefore be 2.7 x 0.88 = 2.4.

If you are not sure whether you need to make an absorption correction, work out your analysis without an absorption correction, guess a likely foil thickness (some previous experience in actually measuring foil thicknesses will of course be necessary here) and see whether an absorption correction using this foil thickness makes any difference to the answer you obtain. If it does, within the limits of accuracy you have set yourself, you will need to measure the foil thickness and make the correction properly.

It is a good idea in any case to plot a graph of I_A/I_B against specimen thickness. This can then be extrapolated to zero thickness to give the thin specimen value of the X-ray intensity ratio. This method has the advantage of revealing and avoiding the effect of any polishing film on the surfaces of the specimen. Polishing films (thin layers with a composition different from that of the bulk material) are common results of electropolishing and ion beam thinning. They are one of the most important problems with thin specimen microanalysis. This procedure is not as time-consuming as it sounds because a *relative* measure of specimen thickness is all that is required. This can easily be obtained by monitoring the intensity in a 'window' placed somewhere on the bremsstrahlung part of the spectrum. This intensity will obviously be proportional to specimen thickness. Of course, one absolute measure of thickness allows the whole axis to be scaled. The ratio of the absorption coefficients, for example, can then be measured.

Fluorescence correction for transmission specimens

This is less important and less common than the absorption correction. It arises when one of the characteristic X-ray wavelengths fluoresces another. This changes the ratio of the measured intensities from what might have been expected. Fluorescence by the continuum, or background, X-rays can be neglected for transmission specimens, as shown by Twigg.[45] (This is not true for bulk, SEM specimens.)

If the same approach as that described for bulk specimens in section 4.2.2.4 is applied to thin specimens, the following expression[46] may be derived for the fluorescence of an X-ray line belonging to element A by one belonging to element B:

$$\frac{I_A^f}{I_A} = \frac{C_B^m A_A}{A_B} \times \frac{U_B \ln(c_B U_B)}{U_A \ln(c_A U_A)} \times \omega_B \times \frac{r_A - 1}{r_A} \times \left(\frac{\mu}{\rho}\right)_B^A \times \frac{\rho t}{2} \times$$

$$\left[0.923 - \ln\left\{\left(\frac{\mu}{\rho}\right)_B^{spec} \rho t\right\}\right] \quad (4.5)$$

where A and B refer both to the elements and to the particular X-ray lines being used to monitor them.

C_B^m is the mass concentration of B
A_A is the relative atomic mass for A
U_A is the overpotential for the particular A X-ray line
$c_{A,B}$ are defined in equation (2.4) (as $c_{n,l}$) (note *not* concentration)
ω_B is the fluorescence ratio for the B X-rays
$(r_A - 1)/r_A$ is the edge jump ratio for excitation of the A X-rays (see section 2.7)
ρ is the density of the specimen
t is the specimen thickness

and $\left(\frac{\mu}{\rho}\right)_B^A$ and $\left(\frac{\mu}{\rho}\right)_B^{spec}$ are the mass absorption coefficients for the B X-rays by A atoms and the specimen as a whole, respectively.

As for bulk specimens the true intensity of B X-rays in the specimen must either be deduced from the measured intensity, or calculated. In equation (4.5) it has been calculated.

The terms involving t represent an approximation to a logarithmic integral.

Thin film fluorescence is only of importance when the fluorescing X-ray has an energy just above the excitation energy for the fluoresced X-ray. For example for K-K fluorescence this means fluorescence of element Z by element Z + 2 for 20 < Z < 30. The fluorescence correction for thin specimens is of such restricted usefulness that I shall say nothing more about it here. Further details may be obtained from Nockolds *et al*[46] or from Williams.[40]

4.1.2.3 What about *very* thick specimens?

Even for our treatment of moderate thickness specimens, I have assumed that the beam electrons lose no energy and are not scattered out of their original path. Neglect of these two aspects of the beam's interaction with the specimen is what distinguishes transmission microanalysis correction calculations from those for bulk microanalysis. In fact electron scattering causes the beam to spread even in transmission specimens, and so degrades the spatial resolution of the microanalysis from what might have been expected from the size of the electron beam. Because it has nothing specifically to do with working out the chemical composition I have chosen to defer this topic until Chapter 5. We might ask here, however, whether the approximations of neglecting energy loss and electron scattering *insofar as they affect the basic correction procedures*, ever break down for transmission specimens. The answer, of course, depends on what accuracy you are seeking from your analysis, but I am not aware of any instances in the published literature where anyone has used energy loss or electron scattering calculations in a transmission context to improve genuinely a composition calculation (leaving aside speculative theoretical papers). In any case, it is easy enough using the expressions to be developed in section 4.2 to estimate at what thickness you need to worry about such effects. In general it will be low Z elements where any problem arises. The loss of transmission for high Z elements will reduce comparatively the probability of any such problems, despite the increased beam scattering.

4.2 Bulk specimen X-ray analysis[*]

In bulk specimen microanalysis, instead of the beam electrons passing straight through the specimen with negligible energy loss and little spreading of the beam, as in TEM/STEM microanalysis, the electron beam buries itself in the solid specimen, spreading out over an area much larger than the original beam, and all the electrons come to a virtual halt. Thus when we try to calculate microanalysis correction factors we are very much concerned with fast electron energy loss and fast electron scattering, two things we were able to ignore for TEM/STEM microanalysis. A complete *ab initio*

---

[*]Good general references are given in section 4.2.2.

calculation, even a rough one, of the relative X-ray intensities expected from a multicomponent specimen is rather complicated for a bulk specimen and SEM or EPMA microanalysis is generally done using standards. Sometimes, particularly for binary systems, the standards are used in a simple calibrating mode, effectively to create a graph of ratio of X-ray intensities versus ratio of chemical concentrations. For multicomponent systems this clearly becomes rapidly impracticable and serious microanalysis generally involves the so-called ZAF corrections (from Z: atomic number; A : absorption; F : fluorescence).

Just as the use of standards in thin film microanalysis (section 4.1) effectively provides a measure of the relative cross-sections of the elements, so it does in bulk microanalysis. In thin film microanalysis corrections have to be made for absorption and fluorescence for thicker specimens. In bulk specimen microanalysis several corrections always have to be made involving atomic number related, absorptive and fluorescent effects. These constitute the ZAF corrections referred to above. There is one fundamental difference, however, between the ways in which the analyses are usually performed. For thin film TEM/STEM microanalysis it is customary to measure *relative* X-ray intensities and obtain a relative analysis where the concentrations of the analysed elements are arbitrarily made to add up to 100% (or some other appropriate figure). For bulk SEM or EPMA microanalysis it is usual to measure X-ray counts *per beam electron* and hence, by knowing the concentration of the element in the standard, to deduce an *absolute* concentration of that element in the unknown. Thus, it is quite sensible within this scheme to use pure element standards, for instance. This fundamental difference between the ways in which the analyses are usually performed makes no difference to the correction procedures, which still involve the same basic calculations. The reason for the difference in approach is that in a thin foil standard, varying specimen thickness will affect X-ray intensities but not their ratios (to a first approximation). For bulk specimens this problem just does not arise.

Because these are the normal ways of doing thin film and bulk analyses, respectively, this does not mean that the methods cannot be exchanged. Once the thickness has been calculated for a foil, an absolute concentration can be deduced and, similarly, the ratios method can be used in the SEM. For the purposes of simplicity, however, I will assume in the following description of bulk specimen microanalysis that it is *absolute* concentrations which are required.

The principle of the method, then, is that X-ray intensities are measured from the unknown specimen and from a series of standards containing, respectively, all the elements of interest. The current into each sample must either be the same, or at least known. The ZAF calculation then elucidates the chemical composition of the unknown specimen.

As with my description of thin film correction procedures, I will deal first of all with extracting from the recorded spectrum the true characteristic X-ray intensities, but including any attenuating effect of the detector. I will then describe the ZAF correction procedure.

4.2.1 Extracting the X-ray intensities (including any attenuating effect of the detector)

4.2.1.1 EDX detectors

The situation here is identical to that described in section 4.1.1, except that the form of the background should be different. Actually, at a first glance (compare Fig. 3.4(d) with Fig. 4.2 or Fig. 4.5) it does not appear radically different, at least in shape. What is obvious is that it is larger *vis-à-vis* the characteristic peak sizes when compared with a spectrum from a thin specimen in a TEM/STEM. Physically the spectrum is produced by electrons of all energies between the beam energy and zero at increasing depths within the specimen, which leads to absorption of the background by the specimen. Assuming (incorrectly) that there is a one-to-one relationship between electron energy and depth within the specimen and that this relationship is $E^2 \propto$ depth below the surface (Whiddington[16] - see equation (2.9)) Kramers[27] suggested that the background for a thin specimen, equation (2.15), should, for a thick specimen, be modified to

$$\frac{d\sigma_{br}}{dE} \propto \frac{Z(E_0 - E)}{E} \qquad (4.6a)$$

This can be improved by using a more realistic expression for electron energy vs depth below the specimen surface. The expression used by Fiori *et al.*[47] and subsequently in the NBS FRAME C computer program,[48] had

$$\frac{d\sigma_{br}}{dE} \propto \frac{Z}{E} \left\{ k_1(E_0 - E) + k_2(E_0 - E)^2 \right\} \qquad (4.6b)$$

where k_1 and k_2 are fitting constants.

There is substantial absorption of the bremsstrahlung in the specimen itself and a factor is introduced to account for this. This is normally taken to be the same as that used for the absorption in the specimen of characteristic X-rays. (An absorption correction for characteristic X-rays is derived in section 4.2.2.3 : equation (4.18).) In fact bremsstrahlung is generated rather more deeply in the specimen than characteristic X-rays, but the error involved tends to cancel with errors connected with the generation of the bremsstrahlung. As with the background to thin specimen X-ray spectra, the expression must now be multiplied by the response function of the detector. A larger proportion of the observed background in an SEM spectrum originates from the specimen than is the case with a TEM or STEM. This is because there is less hardware in the immediate vicinity of the specimen in the former case.

Running through the development of section 4.1.1; the EDX spectrum background may be stripped either by frequency filtering or by using a model expression such as equation (4.6b) scaled to the spectrum. Escape peaks may then be removed. Finally the stripped peaks, or transforms in the case of the filtering method, must be turned into characteristic X-ray intensities by fitting either to stored peaks/transforms or to computed peaks.

4.2.1.2 WDX detectors

The peak-to-background ratio for an X-ray peak from a WDX detector is many times higher than that for an EDX detector. Background subtraction is therefore far less important and far easier for a WDX spectrum. Fig. 3.9(a) shows a typical section of a WDX spectrum. Clearly, all that is necessary here is to connect either side of the peak with a straight line and then to subtract the area below it. On the rare occasions when a more sophisticated approach might seem worthwhile - for example low energy peaks - the methodology described for EDX detectors (see section 4.2.1.1) may be employed.

Escape peaks and window absorption both affect WDX spectrometers as well as EDX spectrometers. Because, however, WDX spectrometers are always used on SEMs with bulk specimens and because under these circumstances the same X-ray line is measured from sample and standard, this type of correction, affected by the efficiency of the spectrometer, factors out. Even dead time corrections, which link the efficiency of the detector to counting rate and might be expected to affect a comparison of peak sizes from sample and standard, do not really matter for WDX spectrometers because they can cope with much greater X-ray intensities than EDX detectors. Processing of WDX spectra is therefore very simple and straightforward and usually consists solely of linear interpolation of background under peaks and subsequent measurement of their heights.

4.2.2 X-ray intensities → chemical composition (the ZAF corrections)

For bulk specimens the method for deriving chemical compositions from measured X-ray intensities is fundamentally different from that used for thin film transmission specimens. For thin specimens, because the X-ray intensities depend on the varying and often unknown specimen thickness, only *ratios* of compositions are worked out using either standards of known composition or the purely computational standardless approach. For bulk specimens, the measured X-ray intensities from the sample are compared with the intensities from elemental standards or standards of a known composition. This allows the *absolute* concentration of an element in the irradiated volume of material to be deduced.

For EDX detection it is sufficient to store the elemental spectra and merely collect *one* reference spectrum during the actual analysis. This is always from the same specimen. It measures effectively beam current (taking into account as well any movement of the detector towards or away from the specimen) and may also be used to characterise the electronics of the detector (positions and widths of peaks). For WDX detection, however, it is necessary to collect spectra from each of the standards for each new specimen. This is because the mechanically more complicated WDX spectrometer cannot be set up reproducibly to give exactly the same response.

To a first approximation the concentration of an element in the sample of unknown composition is given by the ratio of X-ray intensities between the sample and the pure element (or a compound, *pro rata*). There are several reasons why this is not absolutely (sometimes even approximately) true. These reasons are as follows:

Z   (i)   The different atomic numbers also mean that the electrons slow down at different rates within the standard and unknown specimen - i.e. these have different *stopping powers*.

126 *Chemical Microanalysis Using Electron Beams*

Because the chance of exciting a given characteristic X-ray varies with beam electron energy, this will also give rise to a difference in characteristic X-ray emission.

Z (ii) The element (let us stick to elemental standards for the moment, for simplicity) and the sample will have different atomic numbers (*average* atomic number in the case of the sample). This means that different proportions of the incident beam electrons will be backscattered and the standard and unknown will, by the same token, see differing numbers of electrons.

A (iii) The absorption of the characteristic X-rays by the specimen itself will be different in the unknown sample and in the standard.

F (iv) Excitation or *fluorescence* of the measured characteristic X-rays by the other characteristic X-rays will be different.

F (v) Fluorescence of the measured characteristic X-rays by the bremsstrahlung will also differ.

Of this list of five effects, the first two depend on atomic number, Z, the third involves absorption (A) and the last two fluorescence (F). Hence the corrections necessary to remove these effects are called the ZAF corrections.

Suppose the X-ray intensity measured from an alloy containing an element is $I^{specimen}$ and that the intensity measured from the pure element is $I^{element}$. The first approximation to the concentration C of the element in the specimen

$$C^{specimen} = \frac{I^{specimen}}{I^{element}}$$

where the concentrations are mass* concentrations.

---

*One might imagine a logical choice for C would be in terms of atoms of A/unit volume. For reasons more historical than physical, mass concentration is used instead. Provided the corrections are worked out consistently with the initial choice, this is completely arbitrary.

We know, however, for the five reasons listed above that this equality

$$C^{\text{specimen}} = \frac{I^{\text{specimen}}}{I^{\text{element}}} F \qquad (4.7)$$

does not hold. We therefore write
where $F = F_s F_b F_a F_{ch} F_{co}$

and the suffixes refer to the five corrections:

- s   stopping power
- b   backscattering
- a   absorption
- ch   fluorescence by **characteristic** X-rays
- co   fluorescence by **continuum** X-rays

The extension to the case where the standard is not an element is obvious.

I will first of all describe each of the five ZAF corrections in turn (sections 4.2.2.1-5) and then give an example of their joint application. The corrections for thin film specimens described earlier in this chapter followed fairly directly from the theory of Chapter 2. Unfortunately, for a bulk specimen the situation is much more complicated and the corrections are largely empirically based, although their *form* is controlled by the expressions developed in Chapter 2. For each of the corrections there is a wide choice of approaches available. I merely present one popular choice for each of these. For those who wish to read more deeply I provide page references to three standard texts: Reed[18], Heinrich[48] and Russ.[30] I suggest these are consulted in the order Russ→Reed→Heinrich. This is in order of readability but inverse order of comprehensiveness. To discuss each of the corrections in turn, as I do, implies that each of them may be applied separately, which is indeed what is usually done. This is a physical nonsense, of course, in reality. All the corrections are intimately connected.

128 *Chemical Microanalysis Using Electron Beams*

Some concession is made to this by *iterating* the five corrections until a consistent answer is obtained. I will return to this after describing each of the five corrections.

4.2.2.1 The stopping power correction

(See Reed[18] Ch. 13; Heinrich[48] section 9 : 2; Russ[30] p. 42.)

When I described corrections for thin film microanalysis I was able to ignore energy losses and scattering of the beam electrons. For bulk microanalysis these are all-important. Instead of the fast electrons passing straight through the microanalysis volume effectively at the same energy, each beam electron entering a bulk specimen is either brought to a virtual halt or re-emitted from the surface of the specimen. In this subsection I shall describe the effect of the first of these processes on the overall ionisation of the atoms in the specimen, and in the next subsection (4.2.2.2) I shall deal with the loss of beam electrons from the specimen.

If you look back to Chapter 2, section 2.3, and in particular to equation (2.4), you will recall that the probability of exciting a particular characteristic X-ray depends on the energy of the beam electron. Replacing the original beam energy E_o by the continuous variable E and rewriting

$$\sigma_{nl} = \frac{\pi e^4 z_{nl}}{(4\pi\varepsilon_0)^2 E_{nl}^2} \frac{b_{nl}\ln(c_{nl}U_{nl})}{U_{nl}} \qquad (4.8)$$

equation (2.4) in terms of the overpotential $U_{nl} = E/E_{nl}$,

As the beam electrons slow down inside the specimen their potential for exciting the nl shell electrons rises to a maximum (at $U_{nl} = e/c_{nl}$) and then falls away to zero as E approaches E_{nl}. The way in which the energy E of the electron decreases to the 'ambient' level inside the solid does not depend very much on the relatively rare collisions with the strongly bound inner orbital electrons, which give rise to the characteristic X-rays we use for microanalysis. Rather, most of the energy loss comes from multiple collisions with the outer, valence electrons. The expression derived by Bethe[5] to describe this slowing down, or energy loss, was presented in Chapter 2, section 2.3:

$$-\frac{dE}{ds} = \frac{\pi e^4 Z}{(4\pi\varepsilon_0)^2 E} N^v 2 \ln\left(\frac{1.166 E}{J}\right) \qquad (2.8)$$

where s is path length, Z atomic number, N^v atoms/unit volume and J the mean excitation energy. For the same reason as with X-ray absorption coefficients (section 2.6), *mass thickness*, ρs, is used instead of distance s. $-dE/d(\rho s)$ is called the *stopping power* S of the material and from equation

$$S(E) = \frac{2\pi e^4}{(4\pi\varepsilon_0)^2} N^m \frac{Z}{E} \ln\left(\frac{1.166 E}{J}\right) \quad J m^2 kg^{-1}$$

$$= 7846 \frac{Z}{AE} \ln\left(\frac{1.166 E}{J}\right) \qquad keV\, m^2 kg^{-1} \qquad (4.9)$$

(2.8) above:
N^m is atoms/unit mass. A is relative atomic mass. The units of S look a little strange because mass thickness is being used instead of distance. (Note two unconnected uses of symbol 'J'.)

The stopping powers of the sample and standard will be different and therefore the respective efficiencies of each beam electron in producing characteristic X-rays, integrated over its total life from entry into the solid down to the ionisation energy, will be different. It is the purpose of the stopping power correction to account for and thereby remove this difference.

If the cross-section per atom for a particular interaction is σ and there are N^v atoms per unit volume, then there will be σN^v collisions per unit length (a cylinder of unit length and cross-sectional area σ contains σN^v atoms). The total number of nl ionisations integrated over the trajectory of the beam

$$n_{nl} = \int_{E_0}^{E_{nl}} \sigma_{nl} N^v \frac{ds}{dE} dE$$

electron will then be

130 *Chemical Microanalysis Using Electron Beams*

In terms of the stopping power S (= $-dE/d(\rho s)$) the total number of nl

$$n_{nl} = N^m \int_{E_{nl}}^{E_0} \frac{\sigma_{nl}}{S} dE \qquad (4.10)$$

ionisations per beam electron
where N^m = atoms/unit mass. Substituting equations (4.8) and (4.9) for σ_{nl}

$$n_{nl} = N^m \int_{E_{nl}}^{E_0} \frac{\pi e^4 z_{nl}}{(4\pi\varepsilon_0)^2 E_{nl}^2} b_{nl} \frac{\ln(c_{nl} U_{nl}) (4\pi\varepsilon_0)^2 E}{U_{nl} 2\pi e^4 N^m Z \ln\left(\frac{1.166 E}{J}\right)} dE$$

$$= \frac{z_{nl} b_{nl}}{2Z} \int_{1}^{U_{nl}^0} \frac{\ln(c_{nl} U_{nl})}{\ln\left(\frac{1.166 E_{nl} U_{nl}}{J}\right)} dU_{nl}$$

and S into equation (4.10),
where $U_{nl}^0 = \frac{E_0}{E_{nl}}$ and z_{nl} is the number of electrons in the nl shell.

Writing $1.166 E_{nl}/J$ as X_{nl}, the integral above can be written in terms of the 'logarithmic integral' Li where

$$Li(x) = \int_0^x \frac{dt}{\ln t}$$

Thus:

$$n_{nl}^{\text{element}} = \frac{z_{nl} b_{nl}}{2Z} \left[U_{nl}^0 - 1 - \frac{\ln\left(\frac{X_{nl}}{c_{nl}}\right)}{X_{nl}} \left\{ \text{Li}\left(X_{nl} U_{nl}^0\right) - \text{Li}(X_{nl}) \right\} \right] \quad \begin{array}{l} \text{nl ionisations} \\ \text{per incident} \\ \text{electron} \end{array}$$

$$\ldots (4.11)$$

The calculation of Li (x) is performed conveniently by computer. To enable calculation by hand here, I have included a table of Li (x) as Table A.7, Appendix A. J in this expression is taken to be 11.5Z eV (Wilson[49]) or, for Z > 12, $(9.76 + 58.5Z^{-1.19})$ Z eV (Berger and Seltzer[50]).

Although I have written equation (4.11) in a general form involving b_{nl} and c_{nl}, it is important to realise that the choice of c_{nl} is not as arbitrary as heretofore. This is because the original integral is from $U_{nl} = 1$ to $U_{nl} = U_{nl}^0$. Any choice of cross-section for which $c_{nl} < 1$ will result in negative cross-sections as U approaches 1. Of the various combinations of b_K and c_K presented in section 2.3, only that due to Green and Cosslett[9] ($b_K = 0.61$; $c_K = 1$) satisfies this criterion. I shall therefore always use Green and Cosslett parameters for K stopping power calculations. (In fairness to the other authors, their parameterisations were not intended for use in the low U regime.) Obviously the same criterion must be applied to L shell stopping power corrections.

For a compound specimen N_i^m, the number of i atoms per unit mass, becomes $N_i^m C_i^m$. Remembering that our correction procedure is conventionally based on *mass* concentrations, equation 4.7, this is very convenient. For similar reasons stopping power is also compounded via mass concentrations:

$$S = \sum_i S_i C_i^m \qquad \text{for the i elements}$$

or, from equation (4.9)

$$S = \frac{2\pi e^4}{(4\pi\varepsilon_0)^2} \frac{1}{E} \sum_i C_i^m N_i^m Z_i \ln\left(\frac{1.166 E}{J_i}\right)$$

The overall form of equation (4.9) can be retained if we set

$$\bar{Z} = \sum_i C_i^a Z_i$$

$$\ln \bar{X} = \frac{\sum_i C_i^a Z_i \ln Z_i}{\bar{Z}}$$

.... (4.12)

where C_i^a is the atomic concentration of element 'i' ($= C_i^m N_i^m$). (Remember $\bar{X}_i$ is specific to a particular X-ray wavelength.) These averaged values should be used in equation (4.11):

$$n_{nl}^{specimen} = C_i^a \frac{z_{nl} b_{nl}}{2\bar{Z}} \left[U_{nl}^0 - 1 - \frac{\ln\left(\frac{\bar{X}_{nl}}{C_{nl}}\right)}{\bar{X}_{nl}} \left\{ Li(\bar{X}_{nl} U_{nl}^0) - Li(\bar{X}_{nl}) \right\} \right] \text{nl ionisations per incident electron}$$

.... (4.11a)

Then
$$F_s = \frac{n_{nl}^{element}}{n_{nl}^{specimen}} C_i^m \quad \text{(see equation (4.7))}$$

Equations (4.11) and (4.12) are based on the Bethe stopping power expression. A simpler, earlier, expression due to Whiddington[16] was mentioned in Chapter 2, equation (2.9). Using this in place of the Bethe expression results in an alternative version of equation (4.11):

$$n_{nl}^{element} = 0.26 \frac{z_{nl} b_{nl}}{A} \left\{ U_0 \ln U_0 + (1 - \ln c_{nl})(1 - U_0) \right\} \quad (4.11b)$$

which was further simplified for K X-rays by Green and Cosslett[9] to

$$n_{nl}^{\text{element}} = \frac{0.116}{A} (U_0 - 1)^{1.67} \qquad (4.11c)$$

Very little is lost by using (4.11b) or (c) in place of (4.11) and thus saving the use of the logarithmic integral, but for the sake of uniformity I will retain equations (4.11)/(4.11a) and (4.12).

Example calculation 4.5

What is the stopping power correction for CuK_a X-rays in cupric sulphide (CuS) with pure copper as standard? (Beam voltage = 25 kV.)

Recall from equation (4.11a) that

$$n_K = C_i^a \frac{z_K b_K}{2\overline{Z}} \left[U_K^0 - 1 - \frac{\ln\left(\frac{\overline{X}_K}{C_K}\right)}{\overline{X}_K} \left\{ \text{Li}(\overline{X}_K U_K^0) - \text{Li}(\overline{X}_K) \right\} \right] \quad \text{K ionisations per incident electron}$$

where
$$\overline{Z} = \sum_i C_i^a Z_i$$

and
$$\ln \overline{X}_K = \frac{\sum_i C_i^a Z_i \ln X_K^i}{\overline{Z}} \qquad \Bigg\} \quad \ldots (4.12)$$

$$X_K^i = 1.166 \frac{E_K}{J_i} \quad ; \quad J_i = \left(9.76 + 58.5\, Z^{-1.19}\right) Z \quad (Z > 12)$$

From Appendix A, Tables A.2, A.6

$$Z_{Cu} = 29 \,; \quad J_{Cu} = \left(9.76 + 58.5 \times 29^{-1.19}\right) 29$$

$$= 314 \text{ eV}$$

$$X_{Cu} = \frac{1.166 \times 8979}{314} = 33.34$$

$$Z_S = 16; \quad J_S = (9.76 + 58.5 \times 16^{-1.19})\,16$$

$$= 191 \text{ eV}$$

$$X_S = \frac{1.166 \times 8979}{191}$$

$$= 54.81$$

For CuS,
$$\overline{Z} = \frac{29 + 16}{2} = 22.5$$

$$\ln \overline{X} = \frac{0.5 \times 29 \times \ln 33.34 + 0.5 \times 16 \times \ln 54.81}{22.5}$$

$$= 3.684$$

$$\therefore \overline{X} = 39.79$$

$$U_K^0 = \frac{25000}{8979} = 2.784$$

$$C_{Cu}^m = \frac{63.55}{63.55 + 32.07}$$

$$= 0.6646$$

Using the Green and Cosslett[9] values for b_K and c_K:

$$b_K = 0.61$$
$$c_K = 1.00$$

Working out the Compositions   135

then for pure copper, using Table A.7 of Appendix A,

$$n^{Cu}_{CuK_\alpha} = \frac{2 \times 0.61}{2 \times 29}\left[2.784 - 1 - \frac{\ln 33.34}{33.34}\left\{Li(33.34 \times 2.784) - Li(33.34)\right\}\right]$$

$$= 0.02103\left\{1.784 - 0.1052\left[27.98 - 13.41\right]\right\}$$

$$= 5.28 \times 10^{-3} \quad \text{ionisations per electron}$$

For CuS,

$$n^{CuS}_{CuK_\alpha} = C^a_{Cu}\frac{2 \times 0.61}{2 \times 22.5} \times$$

$$\left[2.784 - 1 - \frac{\ln 39.79}{39.79}\left\{Li(39.79 \times 2.784) - Li(39.79)\right\}\right]$$

$$= 0.5 \times 0.02711\left[1.784 - 0.09258\left\{31.87 - 15.21\right\}\right]$$

$$= 3.28 \times 10^{-3} \quad \text{ionisations per electron}$$

From equation (4.11a)

$$F_s = \frac{n^{Cu}_{CuK_\alpha}}{n^{CuS}_{CuK_\alpha}} C^m_{Cu} = \frac{5.28}{3.28} \times 0.6646$$

$$= 1.070$$

4.2.2.2. The backscattering correction

(See Reed[18] Ch. 14; Heinrich[48] section 9.3; Russ [30] p. 49.)

Fig. 4.10 shows schematically the shape of the volume of specimen irradiated by the electron beam. (See also Chapter 5, Fig. 5.1.) In particular, some of the beam electrons leave the specimen before their energy has dropped below the excitation energy E_{nl} for the characteristic X-ray of interest. Therefore the efficiency of production of the X-rays is less than might be expected. This is described by a factor R:

$$R = \frac{\text{Number of characteristic X-rays obtained}}{\text{Number of characteristic X-rays which would have been obtained if none of the beam electrons had left the specimen}}$$

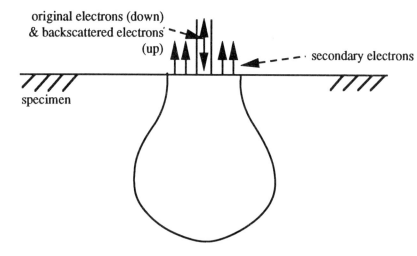

Fig. 4.10 The spreading of an electron beam within a bulk specimen. Some of the electrons are backscattered by the specimen and therefore excite fewer or no characteristic X-rays (see also Fig. 5.1).

Working out the Compositions   137

You will guess from the rather fantastical description of the denominator that R is something which must be calculated rather than measured. The experimental input comes from measurements of the energy distribution of the backscattered electrons. The remainder of the calculation of R lies in predicting what the efficiency of these 'lost' electrons in exciting X-rays would have been. This type of 'stopping-power' calculation was treated in the previous section (4.2.2.1).

Backscattering is caused by large angle scattering of the electrons by the atomic nuclei in the specimen. Large angle elastic scattering was described in section 2.9 and in particular in equations (2.16) and (2.17). From these it is evident that backscattering will increase, and therefore R will decrease, with atomic number. The dependence on beam energy is not so obvious, because it is tied in so much with stopping power. Although the total number of electrons backscattered depends very little on beam energy E, the backscattering correction R does. It is normally expressed as a function of Z and U_o (= E_o/E_{nl}). One formulation is that of Duncumb and Reed:[51]

$$R = [1 \; Z' \; Z'^2 \; Z'^3 \; Z'^4 \; Z'^5] \times$$

$$\begin{bmatrix} 1 & 0 & 0 & 0 & 0 & 0 \\ -0.581 & 2.162 & -5.137 & 9.213 & -8.619 & 2.962 \\ -1.609 & -8.298 & 28.79 & -47.74 & 46.54 & -17.68 \\ 5.400 & 19.18 & -75.73 & 120.05 & -110.70 & 41.792 \\ -5.725 & -21.65 & 88.13 & -136.06 & 117.75 & -42.45 \\ 2.095 & 8.947 & -36.51 & 55.69 & -46.08 & 15.85 \end{bmatrix} \begin{bmatrix} 1 \\ U_o^{-1} \\ U_o^{-2} \\ U_o^{-3} \\ U_o^{-4} \\ U_o^{-5} \end{bmatrix} \quad (4.13)$$

where $Z' = Z/10^2$ and $F_b = R^{element}/R^{specimen}$ (see equation (4.7)).

For compound specimens, since high angle scattering depends on the number of protons/unit volume and this is what controls density, it is logical to set

$$R = \sum_i C_i^m R_i \quad (4.14)$$

where i refers to element i with mass concentration C_i^m and backscattering correction (in the elemental form) R_i.

When the specimen is tilted, R falls. The easiest, but not the only, way of taking this into account is to make sure standard and unknown sample both have the same tilt, and to ignore it. Alternatives will be found in the

138 *Chemical Microanalysis Using Electron Beams*

references given at the beginning of this subsection.

To sum up, the measured X-ray intensity from the standard and the sample must be multiplied by the appropriate value of R, calculated via equation (4.13) [and (4.14), if necessary].

Example calculation 4.6

In the specimen of CuS referred to in Example calculation 4.5, by how much are the SK_α X-rays reduced because of electron loss by backscattering? And the CuK_α? (Beam voltage = 25 kV.)

Using equation (4.13), we have R values as follows:

	U_0	Z 16	29
SK_α	25/2.472 = 10.113	0.901	0.815
CuK_α	25/8.979 = 2.784	0.934	0.864

From Appendix A, Table A.2,

$$C_S^m = \frac{32.07}{32.07 + 63.55} = 0.3354$$

and $C_{Cu}^m = 0.6646$

From equation (4.14)

$$R_{SK} = 0.3354 \times 0.901 + 0.6646 \times 0.815$$
$$= 0.844$$
and $R_{CuK} = 0.3354 \times 0.934 + 0.6646 \times 0.864$
$$= 0.887$$

Note that F_b for SK_α is therefore 0.901/0.844 = 1.068
and   for CuK_α is   0.864/0.887 = 0.974.

4.2.2.3 The absorption correction

(See Reed[18] Ch. 15; Heinrich[48] Ch. 10; Russ[30] Ch. 6.)

This is often the most crucial of the five corrections. Recall the shape of the X-ray generation volume, shown schematically in Fig. 4.10. The characteristic X-rays generated deep in the specimen suffer considerable absorption as they travel through the specimen on their way to the detector. Fig. 4.11 shows the geometry in the simple case of normal incidence of the electron beam on the specimen surface. The length of the escape path is obviously $z/\sin\theta$ where z is the depth of a thin layer below the surface and θ is the take-off angle for the detector.

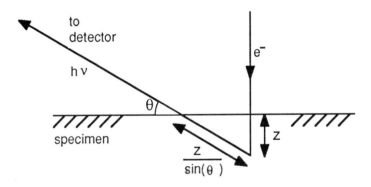

Fig. 4.11 The absorption correction for bulk specimens. The X-ray is generated at depth z and suffers absorption over path length $z/\sin\theta$ on its way out of the specimen to the detector. The angle θ is called the 'take-off' angle of the detector.

As explained in section 2.6, most of the X-ray absorption at the beam voltages of interest here is due to the generation of photoelectrons. It is *mass* absorption coefficients which are tabulated, for reasons given in sections 2.3 and 2.6. Those of Thinh and Leroux[24] are reproduced in Appendix A, Table A.6 and have been used many times previously in this book. A typical energy variation for (μ/ρ) was shown in Fig. 2.10. (μ/ρ) for a multicomponent specimen is given by equation (2.14):

$$\left(\frac{\mu}{\rho}\right) = \sum_i \left(\frac{\mu_i}{\rho_i}\right) C_i^m \qquad (2.14)$$

The absorption of X-rays from the layer shown in Fig. 4.11 is then

$$\exp\left(-\left(\frac{\mu}{\rho}\right)\rho \frac{z}{\sin\theta}\right)$$

Writing $(\mu/\rho)/\sin\theta$ as χ, the absorption factor for a layer at depth z is

$$\exp(-\chi\rho z)$$

If the generation function for the characteristic X-rays for which we are working out the absorption correction is $\phi(\rho z)$, then the total absorption correction is

$$\frac{\int_0^\infty \phi(\rho z) \exp(-\sigma\rho z) \exp(-\chi\rho z) \, d(\rho z)}{\int_0^\infty \phi(\rho z) \, d(\rho z)} \qquad (4.15)$$

where the denominator is a normalising expression for ϕ. Note that the numerator is a Laplace transform of ϕ.

It only remains to derive $\phi(\rho z)$. I will describe the popular approach due to Philibert.[52] Again, this is accomplished by a mixture of simple physical reasoning and subsequent fitting to experimental data. Philibert suggested that the form of $\phi(\rho z)$ is the product of two opposing tendencies: the declining number of electrons below the surface and the increasing path length caused by scattering of the electrons.

Philibert described the declining number of electrons below the surface by Lenard's law, which was introduced in section 2.3, equation (2.10):

$$\text{number of electrons} \quad \alpha \quad \exp(-\sigma\rho z) \qquad (4.16)$$

σ is the Lenard coefficient, the mass absorption coefficient for the electrons. (This should not be confused with the cross-section for ionisation, σ_{nl}.)

Philibert assumed a relative path length factor at the *diffusion depth* (random beam electron directions) of 4 and an exponential decay with an exponent k, giving a factor

$$4 - (4 - \phi(0)) \exp(-k\rho z)$$

Then
$$\phi(\rho z) = \{4 - (4 - \phi(0)) \exp(-k\rho z)\} \exp(-\sigma\rho z)$$

and the absorption factor

$$f_a = \frac{\int_0^\infty \{4 - [4 - \phi(0)] \exp(-k\rho z)\} \exp(-\sigma\rho z) \exp(-\chi\rho z) \, d(\rho z)}{\int_0^\infty \{4 - [4 - \phi(0)] \exp(-k\rho z)\} \exp(-\sigma\rho z) \, d(\rho z)}$$

$$= \frac{\dfrac{1}{\sigma + \chi} - \dfrac{1 - \dfrac{\phi(0)}{4}}{k + \sigma + \chi}}{\dfrac{1}{\sigma} - \dfrac{1 - \dfrac{\phi(0)}{4}}{k + \sigma}}$$

k is usually replaced by σ/h, whereupon

$$f_a = \frac{\dfrac{1}{\sigma + \chi} - \dfrac{1 - \dfrac{\phi(0)}{4}}{\sigma\left(1 + \dfrac{1}{h}\right) + \chi}}{\dfrac{1}{\sigma} - \dfrac{1 - \dfrac{\phi(0)}{4}}{\sigma\left(1 + \dfrac{1}{h}\right)}} \quad (4.17)$$

Unless absorption near the surface is particularly important (i.e. soft X-rays) there is little physical loss and much gain of mathematical simplicity in setting $\phi(0)$ to 0. Then

$$f_a = \frac{1}{\left(1 + \dfrac{\chi}{\sigma}\right)\left(1 + \dfrac{h}{1 + h}\dfrac{\chi}{\sigma}\right)} \quad (4.18)$$

and $F_a = f_a^{element} / f_a^{specimen}$ (see equation (4.7)).

Philibert fitted h and σ to experimental data. Recall $h = \sigma/k$ where σ is the Lenard coefficient; k will be controlled by Rutherford (single atom elastic) scattering. Remembering that the Rutherford cross-section (equation (2.16)) is proportional to Z^2/E^2 at these low SEM energies ($\beta \ll 1$) and that σ is roughly proportional to $1/E^2$, then $h \propto A/Z^2$, where the relative atomic mass A compensates for ρ. Philibert's value of

$$h = 1.2 \, A/Z^2 \qquad (4.19)$$

where A is relative atomic mass and Z atomic number, is still used, but his values for σ have been supplanted by, for example, the expression of Heinrich:[53]

$$\sigma = \frac{4.5 \times 10^4}{E_0^{1.65} - E_{nl}^{1.65}} \quad m^2 \, kg^{-1} \qquad (4.20)$$

where E_0 and E_{nl} are expressed in keV.

It might seem strange that E_{nl} should figure in the electron mass absorption coefficient σ. In fact the $E_{nl}^{1.65}$ accounts for the fact that 'absorption' corresponds to when $E < E_{nl}$. Higher E_{nl} therefore implies higher 'absorption'.

For multicomponent specimens

$$h = \sum_i C_i^m h \qquad (4.21)$$

Example calculation 4.7

To what fraction is the CuK_α X-ray intensity reduced by absorption in a stoichiometric Cu_3Au specimen? The beam voltage is 25 kV and the take-off angle is 40°.

I have used here the example of CuK_α in Cu_3Au because I have already worked out the relevant mass absorption coefficients in Example calculation 2.5. Referring back to this calculation, then, the mass fractions of copper and gold are 0.508 and 0.492 respectively. (μ/ρ) for Cu K_α X-rays of energy 8.04 keV in Cu_3Au was calculated to be 12.86 m² kg⁻¹. For a take-off angle of 40° $\chi = (\mu/\rho)/\sin\theta = 20.01$ m²kg⁻¹.

I am going to use equation (4.18) to calculate the absorption correction. We therefore need, as well as χ, h and σ. h first of all. From equation (4.19), $h_i = 1.2 A_i/Z_i^2$ (note h is dimensionless) and from equation 4.21 $h = \Sigma C_i h_i$. For copper, using Appendix A, Table A.2, $h_{Cu} = 1.2 \times 63.55/29^2 = 0.0907$ and for gold $h_{Au} = 1.2 \times 196.97/79^2 = 0.0379$. Then for Cu_3Au, $h = 0.508 \times 0.0907 + 0.492 \times 0.0379 = 0.0647$.

From equation (4.20), $\sigma = 4.5 \times 10^4/(E_o^{1.65} - E_{nl}^{1.65})$ m² kg⁻¹, with E in keV. Here $E_{CuK} = 8.979$ keV (Appendix A, Table A.6)

$$\sigma = \frac{4.5 \times 10^4}{25^{1.65} - 8.979^{1.65}} = 272.4 \quad m^2 kg^{-1}$$

Substituting in equation (4.18), for CuK_α X-rays in Cu_3Au

$$f_a = \frac{1}{\left(1 + \frac{20.01}{272.4}\right)\left(1 + \frac{0.0647}{1.0647} \times \frac{20.01}{272.4}\right)}$$

$$= 0.927$$

Therefore the intensity of CuK_α X-rays measured at a take-off angle of 40° from Cu_3Au irradiated by 25 kV electrons is reduced by absorption to 0.927 times what it would have been.

Example calculation 4.8

What are the absorption corrections for the K_α X-rays from GaAs (stoichiometric) for a beam voltage of 30 kV and a take-off angle of 40°?

Here I must first work out the absorption coefficients. Ga (Z = 31) K_α X-rays have an energy of 9.241 keV and As (Z = 33) K_α X-rays have an energy 10.542 keV (Appendix A, Table A.3). Then, using the tables of Thinh and Leroux[24] in Appendix A, Table A.6, we have for (μ/ρ) (m² kg⁻¹):

144 *Chemical Microanalysis Using Electron Beams*

Absorbed X-ray	Energy (keV)	Absorber (μ/ρ) (m² kg⁻¹)	
		Ga	As
GaK$_\alpha$	9.241	4.244	5.117
AsK$_\alpha$	10.542	23.784	3.571

Why is one of the mass absorption coefficients so much greater than the other?

Note that the absorption of AsK$_\alpha$ X-rays by Ga atoms is particularly heavy. This is because the AsK$_\alpha$ energy is just above the absorption edge for the GaK, or 1s², shell (10.367 keV). This is significant for fluorescence, as will be seen in the next section 4.2.2.4.

Weight percentages of Ga and As in GaAs are (using the relative atomic masses of Appendix A, Table A.2)

$$C_{Ga}^m = 0.482 \quad \text{and} \quad C_{As}^m = 0.518$$

Then

$$\left(\frac{\mu}{\rho}\right)_{GaK_\alpha}^{GaAs} = 0.482 \times 4.244 + 0.518 \times 5.117$$

$$= 4.696 \text{ m}^2 \text{ kg}^{-1}$$

and

$$\left(\frac{\mu}{\rho}\right)_{AsK_\alpha}^{GaAs} = 0.482 \times 23.784 + 0.518 \times 3.571$$

$$= 13.314 \text{ m}^2 \text{ kg}^{-1}$$

For a 40° take-off angle

for GaK$_\alpha$

$$\chi = \frac{4.696}{\sin 40°} = 7.306 \text{ m}^2 \text{kg}^{-1}$$

and for AsK$_\alpha$

$$\chi = \frac{13.314}{\sin 40°} = 20.713 \text{ m}^2 \text{kg}^{-1}$$

For Ga $\quad h = \dfrac{1.2\, A_{Ga}}{Z_{Ga}^2} = \dfrac{1.2 \times 69.72}{31^2} = 0.0871$

For As $\quad h = \dfrac{1.2\, A_{As}}{Z_{As}^2} = \dfrac{1.2 \times 74.92}{33^2} = 0.0826$

and for GaAs $\quad h = 0.482 \times 0.0871 + 0.518 \times 0.0826$

$\qquad\qquad\qquad = 0.0848 \ \text{m}^2\,\text{kg}^{-1}$

With a beam voltage of 30 kV (equation (4.20))

for $GaK_\alpha \quad \sigma = \dfrac{4.5 \times 10^4}{30^{1.65} - 10.367^{1.65}} = 198.9 \ \text{m}^2\,\text{kg}^{-1}$

and for $AsK_\alpha \quad \sigma = \dfrac{4.5 \times 10^4}{30^{1.65} - 11.867^{1.65}} = 209.9 \ \text{m}^2\,\text{kg}^{-1}$

Then for Ga K_α (equation (4.18))

$$f_a = \dfrac{1}{\left(1 + \dfrac{7.306}{198.9}\right)\left(1 + \dfrac{0.0848}{1.0848} \times \dfrac{7.306}{198.9}\right)}$$

$\qquad = 0.962$

and for As K_α

$$f_a = \dfrac{1}{\left(1 + \dfrac{20.01}{272.4}\right)\left(1 + \dfrac{0.0647}{1.0647} \times \dfrac{20.01}{272.4}\right)}$$

$\qquad = 0.903$

146 *Chemical Microanalysis Using Electron Beams*

The more severe correction for As K_α follows from the heavy absorption of As K_α by Ga. This will be taken up again in Example calculation 4.9.

4.2.2.4 Fluorescence by characteristic X-rays

(See Reed[18] sections 16.1 - 16.8; Heinrich[48] section 11.1; Russ[30] Ch. 7.)

We now come to the fourth of the five ZAF corrections. This relates to X-ray fluorescence by characteristic X-rays. You may recall that X-ray fluorescence is X-ray production where the initial ionising event is accomplished by another X-ray. The physics of the process was described in sections 2.6 and 2.7. The ionisation event, involving the production of photoelectrons, is the dominant process contributing to the X-ray absorption coefficient. The fluorescence and absorption corrections are therefore intimately connected. A large fluorescence correction implies a large absorption correction for one of the X-ray wavelengths. Fluorescence is caused by both characteristic and continuum, or background, X-rays (bremsstrahlung). I shall examine fluorescence by characteristic X-rays here and that by the continuum in the next subsection (4.2.2.5).

Let us take a simple model situation where a specimen consisting of two components, A and B, is being analysed via K X-rays K_A and K_B. (Don't worry for the moment about K_α and K_β.) The energy of the K_B X-rays is higher than that of the K_A X-rays. When we compare the K_A X-ray intensity from the unknown specimen with that from the pure element A, we must ask the question: *what proportion of the K_A X-ray intensity from the unknown specimen is there because it has been fluoresced by the K_B X-rays?* If we are to derive the true concentration of A in the unknown specimen, we must correct for this effect.

The first thing to notice is that the primary X-rays (from electron ionisation of the atoms) come from very close to the specimen surface, but the fluoresced X-rays come from a relatively large volume. Thus for 20 keV electrons, the Lenard coefficient (equation (4.16)) is of the order of hundreds of $m^2 kg^{-1}$, whereas X-ray mass absorption coefficients tend to be an order of magnitude less (with considerable variations, of course).

The situation is depicted in Fig. 4.12. We wish to calculate that part of the K_A intensity, I_A say, due to fluorescence by the K_B intensity, I_B. (It is not important that I have chosen K type X-rays.) Let us for the moment imagine

Working out the Compositions 147

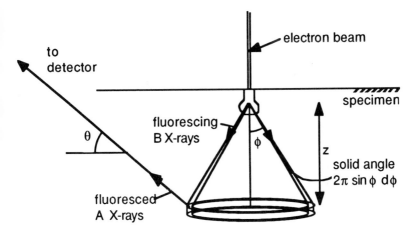

Fig. 4.12 The characteristic fluorescence correction. X-rays from A and B are generated in the electron penetration volume relatively near to the specimen surface. Some of the fluorescing B X-rays head down into the solid. At depth z and azimuthal angle φ having suffered absorption over a distance z/sin φ, the X-rays fluoresce a volume 2π sinφ dφ dz. Those of the resulting A X-rays which are emitted in a direction towards the detector are absorbed over the distance z/sin θ as they escape from the specimen. The positive depth of the electron penetration volume is accounted for via the Lenard coefficient (see text).

I_B being emitted isotropically (perfectly true) from a point source on the specimen surface (not quite true). Half I_B goes into the specimen. Consider a thin annular slice of specimen of thickness dz which subtends a solid angle $2\pi \sin\phi\, d\phi$ to the point source of K_B X-rays. The thin annular slice is therefore irradiated by a K_B intensity

$$I_B \frac{2\pi \sin\phi\, d\phi}{4\pi} \exp\left(-\left(\frac{\mu}{\rho}\right)_B^{spec} \rho z \sec\phi\right)$$

where $(\mu/\rho)_B^{spec}$ is the mass absorption coefficient of the K_B X-rays in the specimen. A fraction

$$\left(\frac{\mu}{\rho}\right)_B^{spec} \rho\, dz\, \sec\phi$$

of these X-rays is absorbed in the annular slice. Of this fraction

$$C_A^m \frac{\left(\frac{\mu}{\rho}\right)_B^A}{\left(\frac{\mu}{\rho}\right)_B^{spec}}$$

corresponds to absorption by A atoms.

(Remember $\left(\frac{\mu}{\rho}\right)_B^{spec} = C_A^m \left(\frac{\mu}{\rho}\right)_B^A + C_B^m \left(\frac{\mu}{\rho}\right)_B^B$.)

And of this fraction $(r_A - 1)/r_A$ corresponds to K shell ionisation, where r_A is the K_A absorption jump ratio, introduced in section 2.7. A fraction ω_A of the ionisations leads to K_A X-rays (rather than Auger electrons). Finally (for the moment!) these K_A X-rays are absorbed on their way out of the specimen yielding a factor

$$\exp\left(-\left(\frac{\mu}{\rho}\right)_A^{spec} \rho z \; \text{cosec}\; \theta\right)$$

where $\left(\frac{\mu}{\rho}\right)_A^{spec}$ is the mass absorption coefficient and θ is the take-off angle, defined in Fig. 4.12. Putting all these factors together, the fluoresced K_A intensity I_A^f is given by

$$I_A^f = \int_{z=0}^{\infty} \int_{\phi=0}^{\frac{\pi}{2}} I_B \frac{2\pi \sin\phi \; d\phi}{4\pi} \exp\left(-\left(\frac{\mu}{\rho}\right)_B^{spec} \rho z \sec\phi\right) \rho \, dz \sec\theta$$

$$\times \frac{C_A^m \left(\frac{\mu}{\rho}\right)_B^A}{\left(\frac{\mu}{\rho}\right)_B^{spec}} \frac{r_A - 1}{r_A} \omega_A \exp\left(-\left(\frac{\mu}{\rho}\right)_A^{spec} \rho z \; \text{cosec}\; \theta\right)$$

$$= I_B \frac{C_A^m \left(\frac{\mu}{\rho}\right)_B^A}{2} \frac{r_A - 1}{r_A} \omega_A \int_0^{\frac{\pi}{2}} \frac{\sin\phi \, d\phi}{\left(\frac{\mu}{\rho}\right)_B^{spec} + \left(\frac{\mu}{\rho}\right)_A^{spec} \csc\theta \cos\phi}$$

$$= I_B \frac{C_A^m}{2} \left(\frac{\mu}{\rho}\right)_B^A \frac{r_A - 1}{r_A} \omega_A \frac{\ln\left\{1 + \frac{\left(\frac{\mu}{\rho}\right)_A^{spec} \csc\theta}{\left(\frac{\mu}{\rho}\right)_B^{spec}}\right\}}{\left(\frac{\mu}{\rho}\right)_A^{spec} \csc\theta} \qquad (4.22)$$

The terms following I_B in the equation above define the efficiency of the specimen for producing fluoresced K_A X-rays from a point source of K_B X-rays.

At this stage you might imagine that since we will have measured I_B from the unknown specimen, equation (4.22) will enable us now to perform the characteristic fluorescence correction. Unfortunately, this is not so, at least not for WDX. The essence of the WDX method is that the *same X-ray wavelength* is monitored from different specimens. This approach means that we can avoid worrying about the variation in efficiency of the WDX spectrometer for different wavelengths, which is likely to be considerable - for example, it is quite normal to have to use more than one reflecting crystal within one analysis. A corollary of this is that I_B/I_A has to be calculated from the composition of the specimen. (For EDX detectors it should be possible, remembering the energy variation of the detector efficiency, to use the measured I_B.) Here, however, I will continue with the classical (WDX) approach. Remember, as with the other corrections, the answer we are looking for enters the calculation: we have to *iterate* until we arrive at the final composition.) I_B/I_A may be calculated using the stopping power and backscattering corrections described in subsections 4.2.2.1 and 4.2.2.2 above and incorporating appropriate fluorescence yields and partition factors. A somewhat simpler approach, however, due to Reed

150 *Chemical Microanalysis Using Electron Beams*

and Long,[54] is usually employed. This uses the stopping power correction of Green and Cosslett[9] which itself incorporates the Green and Cosslett[9] cross-section (section 2.3) and a simple, pre-Bethe law for stopping power due to Whiddington[16] introduced in section 2.3 as equation (2.9). According to this approach,

$$I \propto \omega_K / (A (U_o - 1)^{1.67})$$

(ω_K : fluorescence ratio; A relative atomic mass; $U_o = E_o/E_K$)

Then
$$\frac{I_B}{I_A} = \frac{C_B^m \, \omega_K^B \, A_A \, (U_B^0 - 1)^{1.67}}{C_A^m \, \omega_K^A \, A_B \, (U_A^0 - 1)^{1.67}} \tag{4.23}$$

where I_A is the intensity of K_A X-rays *before* fluorescence by the K_B X-rays.

Substituting equation (4.23) into equation (4.22) leads to the following expression for the fluoresced K_A X-ray intensity:

$$\frac{I_A^f}{I_A} = \frac{C_B^m}{2} \frac{A_A}{A_B} \omega_K^B \left(\frac{U_B^0 - 1}{U_A^0 - 1}\right)^{1.67} \frac{r_A - 1}{r_A} \left(\frac{\mu}{\rho}\right)_B^A \frac{\ln\left\{1 + \frac{\left(\frac{\mu}{\rho}\right)_A^{spec} \csc\theta}{\left(\frac{\mu}{\rho}\right)_B^{spec}}\right\}}{\left(\frac{\mu}{\rho}\right)_A^{spec} \csc\theta}$$

$$\ldots . (4.24)$$

I assumed at the beginning of this derivation that the fluorescing K_B X-rays were generated at a point on the surface of the specimen. In fact they are generated over a depth controlled by the attenuation of the electron intensity, which is described by Lenard's law (equation (2.10)):

$$\phi(\rho z) \propto \exp(-\sigma \rho z)$$

where σ is the mass absorption coefficient (Lenard coefficient) of the electrons. If we incorporate this expression here, equation (4.24) is changed to:

$$\frac{I_A^f}{I_A} = \frac{C_B^m}{2} \frac{A_A}{A_B} \omega_K^B \left(\frac{U_B^0 - 1}{U_A^0 - 1}\right)^{1.67} \frac{r_A - 1}{r_A} \left(\frac{\mu}{\rho}\right)_B^A$$

$$\times \left[\frac{\ln\left\{1 + \frac{\left(\frac{\mu}{\rho}\right)_A^{spec} \csc\theta}{\left(\frac{\mu}{\rho}\right)_B^{spec}}\right\}}{\left(\frac{\mu}{\rho}\right)_A^{spec} \csc\theta} + \frac{\ln\left\{1 + \frac{\sigma}{\left(\frac{\mu}{\rho}\right)_B^{spec}}\right\}}{\sigma}\right] \quad (4.25)$$

and

$$F_{ch} = \frac{\left(1 + \frac{I_A^f}{I_A}\right)^{element}}{\left(1 + \frac{I_A^f}{I_A}\right)^{specimen}} \quad \text{(see equation (4.7))}$$

Equation (4.25) is our final expression for the fluorescence correction. It is for the fluorescence of K X-rays *by* K X-rays. Other combinations involve changes of detail but not of principle. It did not seem appropriate in a book of this length to include fluorescence of or by L X-rays. Details may be found in the general references.

152 *Chemical Microanalysis Using Electron Beams*

Example calculation 4.9

A Ga As specimen is irradiated by 30 kV electrons. What proportion of the GaK_α X-rays is due to fluorescence by the AsK_α X-rays? (The take-off angle was 40°.)

I will use equation (4.25).
The following data from Example calculation 4.8 are useful here:

$$A_{Ga} = 69.72 \qquad A_{As} = 74.92 \qquad C_{As}^m = 0.518$$

$$\left(\frac{\mu}{\rho}\right)_{GaK_\alpha}^{GaAs} = 4.696 \text{ m}^2\text{kg}^{-1} \qquad \left(\frac{\mu}{\rho}\right)_{AsK_\alpha}^{Ga} = 23.784 \text{ m}^2\text{kg}^{-1}$$

$$\left(\frac{\mu}{\rho}\right)_{AsK_\alpha}^{GaAs} = 13.314 \text{ m}^2\text{kg}^{-1}$$

For the fluorescing As K_α X-rays $\sigma = 209.9$ m² kg⁻¹

$$E_K^{Ga} = 10.367 \text{ keV} \qquad E_K^{As} = 11.867 \text{ keV}$$

The remaining parameters for equation (4.25) are:

$$\omega_{AsK} = 0.562 \qquad \text{(Appendix A, Table A.4)}$$

$$U_{GaK}^0 = \frac{30}{10.367} = 2.894 \qquad U_{AsK}^0 = \frac{30}{11.867} = 2.528$$

Then
$$\left(\frac{U_{AsK}^0 - 1}{U_{GaK}^0 - 1}\right)^{1.67} = 0.699$$

Using the table of Thinh and Leroux[24] in Appendix A (Table A.6) on the high energy side of the Ga K absorption edge,

Working out the Compositions

$$\left(\frac{\mu}{\rho}\right)_{GaK^+}^{Ga} = 1.4662 \times 10.3671 \left(\frac{12.3981}{10.3671}\right)^{2.85}$$

$$= 24.898 \text{ m}^2 \text{ kg}^{-1}$$

and on the low energy side

$$\left(\frac{\mu}{\rho}\right)_{GaK^-}^{Ga} = 1.4662 \times 1.2977 \left(\frac{12.3981}{10.3671}\right)^{2.73}$$

$$= 3.101 \text{ m}^2 \text{ kg}^{-1}$$

Then
$$\frac{r_{GaK} - 1}{r_{GaK}} = \frac{24.898 - 3.101}{24.898}$$

$$= 0.875$$

Substituting into equation (4.25)

$$\frac{I_{GaK\alpha}^f}{I_{GaK\alpha}} = \frac{C_{As}^m}{2} \frac{A_{Ga}}{A_{As}} \omega_K^{As} \left(\frac{U_{AsK\alpha}^0 - 1}{U_{GaK\alpha}^0 - 1}\right)^{1.67} \frac{r_{GaK} - 1}{r_{GaK}} \left(\frac{\mu}{\rho}\right)_{AsK\alpha}^{Ga}$$

$$\times \left[\frac{\ln\left\{1 + \dfrac{\left(\dfrac{\mu}{\rho}\right)_{GaK_\alpha}^{GaAs} \cosec \theta}{\left(\dfrac{\mu}{\rho}\right)_{AsK_\alpha}^{GaAs}}\right\}}{\left(\dfrac{\mu}{\rho}\right)_{GaK_\alpha}^{GaAs} \cosec \theta} + \dfrac{\ln\left\{1 + \dfrac{\sigma}{\left(\dfrac{\mu}{\rho}\right)_{AsK_\alpha}^{GaAs}}\right\}}{\sigma} \right]$$

$$= \frac{0.518}{2} \times \frac{69.72}{74.92} \times 0.562 \times 0.699 \times 0.875 \times 23.784$$

$$\times \left[\frac{\ln\left\{1 + \frac{4.696}{13.314 \times \sin 40°}\right\}}{4.696 / \sin 40°} + \frac{\ln\left\{1 + \frac{209.9}{13.314}\right\}}{209.9} \right]$$

$= 0.144$

The characteristic fluorescence correction is appreciable only when one of the wavelength energies is just above the absorption edge for another of the X-ray energies being monitored. For K-K fluorescence this means that the two respective atomic numbers have to be 1 (low Z) or 2 (medium Z) apart. Clearly the characteristic fluorescence correction is more important when the fluoresc*ing* element occupies a large mass fraction and the fluoresc*ed* element is quite dilute.

4.2.2.5 Fluorescence by continuum X-rays

(See Reed[18] section 16.5; Heinrich[48] section 11.2.)

The continuum fluorescence correction tends to be relatively unimportant not so much because the continuum X-rays do not fluoresce the specimen, but rather because the effects do not vary much from sample to sample and therefore cancel out. The relative unimportance of the correction means that it has received commensurately little attention. Here is a procedure described by Reed:[18]

The continuum X-rays are assumed to be generated at the surface of the specimen (see section 4.2.2.4). The Kramers[27] expression for the overall energy distribution of the continuum (classical theory +Whiddington's[16] stopping power approximation) was given earlier as:

$$\frac{d\sigma_{br}(E)}{dE} \propto \frac{Z(E_0 - E)}{E} \qquad (4.6a)$$

Designating the fluoresced line as K_A, and integrating from E_K^A (the K absorption edge) to E_o, the number of continuum X-rays available for fluorescence is proportional to $Z E_K (U_o \ln U_o - U_o + 1)$. Between any two given absorption edges belonging to the specimen, the same fraction

$$\frac{\left(\frac{\mu}{\rho}\right)_c^A}{\left(\frac{\mu}{\rho}\right)_c^{spec}}$$

of the continuum is absorbed by the A atoms in the specimen. Recall that this fraction does not vary much (see Example calculation 2.7) between two neighbouring edges. As in section 4.2.2.4 we have to multiply by ω_K and $(r_A-1)/r_A$. Incorporating ω_K in a constant, we have that

$$\frac{I_A^f}{I_A} \propto \frac{r_A - 1}{r_A} \bar{Z} E_K^A \frac{\left(\frac{\mu}{\rho}\right)_{K^+}^A}{\left(\frac{\mu}{\rho}\right)_{K^+}^{spec}} (U_0 \ln U_0 - U_0 + 1) \qquad (4.26)$$

$$(\bar{Z} = \sum_i^m C_i Z_i)$$

Absorption is taken into account in a way similar to that of section 4.2.2.4, with the complication that it must be incorporated in the original energy integration. Assuming $\mu \propto E^{-3}$ (section 2.6), and approximating the resulting integral, results in the $(U_o \ln U_o - U_o + 1)$ term in equation (4.26) being replaced by

$$\frac{\ln(1 + uU_0)}{uU_0}$$

where $u = \dfrac{\left(\frac{\mu}{\rho}\right)_A^{spec} \csc \theta}{\left(\frac{\mu}{\rho}\right)_{K^+}^{spec}}$ and the numerator refers to the absorption of the fluoresced AK_α X-rays and the denominator to the fluorescing bremsstrahlung, just above the K absorption edge for A.

Then

$$\frac{I_A^f}{I_A} = 4.34 \times 10^{-6} \frac{r_A - 1}{r_A} A_A \bar{Z} E_K^A \frac{\left(\frac{\mu}{\rho}\right)_{AK^+}^A}{\left(\frac{\mu}{\rho}\right)_{AK^+}^{spec}} \frac{\ln(1 + uU_0)}{uU_0} \quad (4.27)$$

and

$$F_{co} = \frac{\left(1 + \frac{I_A^f}{I_A}\right)^{element}}{\left(1 + \frac{I_A^f}{I_A}\right)^{specimen}} \quad \text{(see equation 4.7)}$$

where the constant is due to Springer.[55] (Notice that A_A, the relative atomic mass, has appeared.) E_K^A is in keV and $\left(\frac{\mu}{\rho}\right)_{AK^+}^A$ and $\left(\frac{\mu}{\rho}\right)_{AK^+}^{spec}$ are evaluated just above the absorption edge for A.

If there are other absorption edges between that for K_A, and E_o, then the last three factors in equation (4.27) must be altered. For a binary alloy containing A and B, for example, and where K_B is at a higher energy than K_A,

$$E_K^A \frac{\left(\frac{\mu}{\rho}\right)_{AK^+}^A}{\left(\frac{\mu}{\rho}\right)_{AK^+}^{AB}} \frac{\ln(1 + uU_0)}{uU_0} \qquad \text{is replaced by}$$

Working out the Compositions 157

$$\left[\frac{\left(\frac{\mu}{\rho}\right)^A_{AK^+}}{\left(\frac{\mu}{\rho}\right)^{spec}_{AK^+}}\left\{E^A_K f^A(U_0)\frac{\ln(1+u_A U^0_A)}{u_A U^0_A}-E^B_K f^B(U_0)\frac{\ln(1+u_B U^0_B)}{u_B U^0_B}\right\}\right.$$

$$\left.+\frac{\left(\frac{\mu}{\rho}\right)^A_{BK^+}}{\left(\frac{\mu}{\rho}\right)^{spec}_{BK^+}}\frac{E^B_K f^B(U_0)\ln(1+u_B U^0_B)}{u_B U^0_B}\right]\Bigg/ f^A(U_0) \quad (4.28)$$

where $f^A_A(U_o) = U_o \ln U_o - U_o + 1$ is worked out at the absorption edge for K_A (E^+_K).

I hope equation (4.28) is for the most part instinctively correct to you: if not, I suggest you consult Reed[18] section 16.5. Such a generally minor correction does not deserve more space here.

Example calculation 4.10

For a sample of Ga As irradiated using 30 kV electrons with an X-ray take-off angle of 40°, what fractions of the K_α X-ray yields are due to fluorescence by bremsstrahlung?

Here we use equations (4.27) and (4.28).
From Table A.2, Appendix A,

$$A_{Ga} = 69.72 \qquad A_{As} = 74.92$$

and 50 at % Ga - 50 at % As = 48.20 wt % Ga - 51.80 wt % As
From Table A.6, Appendix A,

$$E_{GaK} = 10.367 \text{ keV} \qquad E_{AsK} = 11.867 \text{ keV}$$

158 *Chemical Microanalysis Using Electron Beams*

$$U^0_{GaK} = 2.894 \qquad U^0_{AsK} = 2.128$$

For AsK_α X-rays, using the Thinh and Leroux[24] data in Appendix A, Table A.6,

$$\left(\frac{\mu}{\rho}\right)^{As}_{AsK_\alpha} = 1.562 \qquad \left(\frac{\mu}{\rho}\right)^{Ga}_{AsK_\alpha} = 23.859 \text{ m}^2\text{kg}^{-1}$$

and

$$\left(\frac{\mu}{\rho}\right)^{GaAs}_{AsK_\alpha} = 1.562 \times 0.518 + 23.859 \times 0.482$$

$$= 12.309 \text{ m}^2\text{kg}^{-1}$$

$$\left(\frac{\mu}{\rho}\right)^{As}_{AsK^+} = 22.751 \qquad \left(\frac{\mu}{\rho}\right)^{Ga}_{AsK^+} = 19.357 \text{ m}^2\text{kg}^{-1}$$

and

$$\left(\frac{\mu}{\rho}\right)^{GaAs}_{AsK^+} = 21.115 \text{ m}^2\text{kg}^{-1}$$

$$\left(\frac{\mu}{\rho}\right)^{As}_{AsK} = 2.908 \text{ m}^2\text{kg}^{-1} \quad ; \quad \frac{r_{AsK} - 1}{r_{AsK}} = 0.8722$$

$$u = \frac{\left(\frac{\mu}{\rho}\right)^{GaAs}_{AsK_\alpha}}{\left(\frac{\mu}{\rho}\right)^{GaAs}_{AsK^+} \sin\theta} = \frac{12.309}{21.115 \sin 40^\circ} = 0.9069 \; ; \; uU_0 = 2.293$$

$$\overline{Z} = 0.482 \times 31 + 0.518 \times 33 = 32.04$$

Working out the Compositions 159

From equation (4.27)

$$\frac{I_{AsK_\alpha}^f}{I_{AsK_\alpha}} = 4.34 \times 10^{-6} \times 0.8722 \times 74.92 \times 32.04 \times 11.867$$

$$\times \frac{22.751}{21.115} \frac{\ln(1 + 2.293)}{2.293}$$

$$= 0.060$$

For Ga K_α X-rays we have to consider two ranges: one from the Ga K absorption edge up to the As K absorption edge and the other from the As K absorption edge up to E_o, the initial energy of the electrons and therefore the highest possible energy for the bremsstrahlung.

$$\left(\frac{\mu}{\rho}\right)_{GaK_\alpha}^{Ga} = 4.244 \ ; \ \left(\frac{\mu}{\rho}\right)_{GaK_\alpha}^{As} = 5.117 \ ; \ \left(\frac{\mu}{\rho}\right)_{GaK_\alpha}^{GaAs} = 4.696 \ m^2 kg^{-1}$$

$$\left(\frac{\mu}{\rho}\right)_{GaK^+}^{Ga} = 24.898 \ ; \ \left(\frac{\mu}{\rho}\right)_{GaK^+}^{As} = 3.738 \ ; \ \left(\frac{\mu}{\rho}\right)_{GaK^+}^{GaAs} = 13.937 \ m^2 kg^{-1}$$

$$\left(\frac{\mu}{\rho}\right)_{GaK^-}^{Ga} = 3.101 \ m^2 kg^{-1} \ ; \ \frac{r_{GaK} - 1}{r_{GaK}} = 0.8755$$

$$u_{Ga} = \frac{\left(\frac{\mu}{\rho}\right)_{GaK_\alpha}^{GaAs}}{\left(\frac{\mu}{\rho}\right)_{GaK^+}^{GaAs} \sin 40°} = \frac{4.696}{13.937 \sin 40°} = 0.5242$$

$$u_{As} = 0.9069$$

$$\overset{Ga}{f}(U_0) = 2.894 \ln 2.894 - 2.894 + 1$$
$$= 1.181$$

$$\overset{As}{f}(U_0) = 2.528 \ln 2.528 - 2.528 + 1$$
$$= 0.817$$

160 *Chemical Microanalysis Using Electron Beams*

Noting that

$$u_{Ga} U^0_{GaK} = 1.517 \quad \text{and} \quad u_{As} U^0_{AsK} = 2.293$$

and substituting into equation (4.28), this absorption part of the correction for continuum fluorescence of Ga K_α X-rays in GaAs becomes

$$\left[\frac{24.898}{13.937} \left\{ \begin{array}{l} 10.367 \times 1.181 \times \frac{\ln(1+1.517)}{1.517} \\ -11.867 \times 0.817 \times \frac{\ln(1+2.293)}{2.293} \end{array} \right\} \right.$$

$$\left. + \frac{19.357}{21.115} \times 11.867 \times 0.817 \times \frac{\ln(1+2.293)}{2.293} \right] \Big/ 1.181$$

$$= 7.558$$

Then $\quad \dfrac{I^f_{GaK_\alpha}}{I_{GaK_\alpha}} = 4.34 \times 10^{-6} \times 0.8755 \times 69.72 \times 32.04 \times 7.558$

$$= 0.064$$

4.2.2.6 Summary of formulae used for ZAF corrections

Correction	Equations
Stopping power (s)	4.11, 4.12
Backscattering (b)	4.13
Absorption (a)	4.18 - 4.21
Characteristic fluorescence (ch)	4.25
Continuum fluorescence (co)	4.27, 4.28

4.2.2.7 An example of a full calculation

Example calculation 4.11

The following numbers of X-rays were counted over the same period from a specimen containing iron and nickel and from two elemental standards using a WDX spectrometer. The background has already been subtracted:

	Unknown	Element
Fe K_α	31906	106602
Ni K_α	72062	103853

The beam voltage was 20 kV and the X-ray take-off angle was 40°.

The uncorrected compositions are:

$$C_{Fe}^m = \frac{31906}{106602} = 29.9 \text{ wt \%}$$

$$C_{Ni}^m = \frac{72062}{103853} = 69.4 \text{ wt \%}$$

Our first guess is therefore $\text{Ni} - \frac{29.9}{29.9 + 69.4}$ wt % Fe

$$= \text{Ni} - 30.0 \text{ wt \% Fe}^*$$

I will now run through the five ZZAFF corrections in the same order as earlier.

---

*Alternatively I could have left the two weight percentages as they were.

162 *Chemical Microanalysis Using Electron Beams*

Stopping power   (Equations (4.11), (4.12) and Appendix A, Tables A.2, A.6 and A.7)

Fe $Z = 26$; $J = 285$ eV ; $E_K = 7.112$ keV ; $X = 29.10$; $U_K^0 = 2.812$; $C_{Fe}^a = 0.3105$

Ni $Z = 28$; $J = 304$ eV ; $E_K = 8.333$ keV ; $X = 31.96$; $U_K^0 = 2.400$; $C_{Ni}^a = 0.6895$

For Ni - 30 wt % Fe, $\bar{Z} = 27.38$

For Fe K $\overline{\ln X} = 3.325$ and $\bar{X} = 27.80$

For Ni K $\overline{\ln X} = 3.484$ and $\bar{X} = 32.59$

Then for Fe K_α in the elemental Fe,

$$n_{FeK}^{Fe} = \frac{2 \times 0.61}{2 \times 26}\left[2.812 - 1 - \frac{\ln(29.1)}{29.1}\left\{Li(29.1 \times 2.812) - Li(29.1)\right\}\right]$$

$$= 6.283 \times 10^{-3} \text{ FeK ionisations per electron}$$

Similarly $n_{FeK}^{specimen} = 1.860 \times 10^{-3}$

$n_{NiK}^{Ni} = 3.711 \times 10^{-3}$ NiK ionisations per electron

and $n_{NiK}^{specimen} = 2.609 \times 10^{-3}$

Thus $F_s(FeK_\alpha) = \dfrac{n^{Fe}}{\dfrac{n^{spec}}{0.3}} = 1.013$

and $F_s(NiK_\alpha) = 0.996$

Backscattering (Equations (4.13) and (4.14))

Values of R from equation (4.13):

	U_K^0	Z	
		26	28
Fe K_α	2.812	0.879	0.869
Ni K_α	2.400	0.890	0.880

$$R_{FeK_\alpha}^{specimen} = 0.872 \; ; \; R_{NiK_\alpha}^{specimen} = 0.883$$

$$F_b(FeK_\alpha) = \frac{0.879}{0.872} = 1.008$$

$$F_b(NiK_\alpha) = 0.997$$

Absorption (Equations (2.9), (4.18) - (4.20) and Appendix A, Tables A.2, A.3 and A.6)

Values of $\left(\frac{\mu}{\rho}\right)$ (and χ in parentheses) ($m^2 kg^{-1}$):

	Absorber	
	Fe	Ni
Fe K_α at 6.398 keV	7.023 (10.926)	8.630 (13.426)
Ni K_α at 7.471 keV	38.708 (60.219)	5.652 (8.793)

$$\left(\frac{\mu}{\rho}\right)^{\text{specimen}}_{\text{FeK}_\alpha} = 0.7 \times 8.630 + 0.3 \times 7.023 = 8.148 \, (12.676) \; \text{m}^2\text{kg}^{-1}$$

$$\left(\frac{\mu}{\rho}\right)^{\text{specimen}}_{\text{NiK}_\alpha} = 15.569 \, (24.211) \; \text{m}^2\text{kg}^{-1}$$

$$h_{\text{Fe}} = \frac{1.2 \times 55.85}{26^2} = 0.0991$$

$$h_{\text{Ni}} = 0.0898$$

$$h_{\text{specimen}} = 0.0926$$

$$\sigma(20 \text{ keV, FeK}_\alpha) = \frac{4.5 \times 10^4}{20^{1.65} - 7.112^{1.65}} = 392.2 \; \text{m}^2\text{kg}^{-1}$$

$$\sigma(20 \text{ keV, NiK}_\alpha) = 420.1 \; \text{m}^2\text{kg}^{-1}$$

FeK$_\alpha$ in Fe
$$f_a = \frac{1}{\left(1 + \frac{10.926}{392.2}\right)\left(1 + \frac{0.0991}{1.0991} \times \frac{10.926}{392.2}\right)}$$

$$= 0.970$$

Fe K$_\alpha$ in specimen $f_a = 0.966$
Ni K$_\alpha$ in Ni $f_a = 0.978$
Ni K$_\alpha$ in specimen $f_a = 0.938$

$$F_a(\text{FeK}_\alpha) = \frac{0.970}{0.966} = 1.004$$

$$F_a(\text{NiK}_\alpha) = \frac{0.978}{0.938} = 1.043$$

Working out the Compositions

Characteristic fluorescence (of $Fe\,K_\alpha$ by $Ni\,K_\alpha$) (Equation (4.25) and Appendix A, Tables A.2, A.4 and A.6)

$$\left(\frac{\mu}{\rho}\right)_{FeK_\alpha}^{specimen} = 8.148 \text{ m}^2\text{kg}^{-1} \quad (\chi = 12.676 \text{ m}^2\text{kg}^{-1})$$

$$\left(\frac{\mu}{\rho}\right)_{NiK_\alpha}^{specimen} = 15.569 \text{ m}^2\text{kg}^{-1} \qquad \left(\frac{\mu}{\rho}\right)_{NiK_\alpha}^{Fe} = 38.708 \text{ m}^2\text{kg}^{-1}$$

$$A_{Fe} = 55.85 \quad A_{Ni} = 58.69$$

$$\sigma(20 \text{ kV, Ni K}_\alpha) = 420.1 \text{ m}^2\text{ kg}^{-1}$$

$$\omega_{NiK} = 0.406$$

$$U_{FeK}^0 = 2.812 \qquad U_{NiK}^0 = 2.400$$

$$\left(\frac{U_{NiK}^0 - 1}{U_{FeK}^0 - 1}\right)^{1.67} = 0.6500$$

$$\left(\frac{\mu}{\rho}\right)_{FeK^+}^{Fe} = 44.288 \text{ m}^2\text{kg}^{-1}; \quad \left(\frac{\mu}{\rho}\right)_{FeK^-}^{Fe} = 5.261 \text{ m}^2\text{kg}^{-1}$$

and
$$\frac{r_{FeK} - 1}{r_{FeK}} = 0.8812$$

$$\frac{I_{FeK_\alpha}^f}{I_{FeK_\alpha}} = \frac{0.7}{2} \times \frac{55.85}{58.69} \times 0.406 \times 0.6500 \times 0.8812 \times 38.71$$

$$\times \left\{\frac{\ln\left(1 + \frac{12.676}{15.569}\right)}{12.676} + \frac{\ln\left(1 + \frac{420.1}{15.569}\right)}{420.1}\right\}$$

$$= 0.1647$$

$$\therefore \quad F_{ch}(Fe\,K_\alpha) = 0.859$$
$$(F_{ch}(Ni\,K_\alpha) = 1.000)$$

166 *Chemical Microanalysis Using Electron Beams*

Continuum fluorescence (Equations (4.27) and (4.28) and Appendix A, Tables A.2, A.3 and A.6)

FeK_α in Fe

$$\left(\frac{\mu}{\rho}\right)_{FeK_\alpha}^{Fe} = 7.023 \ (\chi = 10.926) \ m^2 kg^{-1}$$

$$\left(\frac{\mu}{\rho}\right)_{FeK^+}^{Fe} = 44.288 \ ; \quad \left(\frac{\mu}{\rho}\right)_{FeK^-}^{Fe} = 5.261 \ m^2 kg^{-1}$$

$$\frac{r_{FeK} - 1}{r_{FeK}} = 0.8812 \ ; \quad u = \frac{10.926}{44.288} = 0.2467$$

$$\frac{\ln(1 + uU_0)}{uU_0} = 0.7596$$

$$\frac{I_{FeK_\alpha}^f}{I_{FeK_\alpha}} = 4.34 \times 10^{-6} \times 0.8812 \times 55.85 \times 26 \times 7.112 \times \frac{44.288}{44.288} \times 0.7596$$

$$= 0.030 \quad \text{in Fe}$$

FeK_α in specimen

$$\left(\frac{\mu}{\rho}\right)_{FeK_\alpha}^{Ni} = 8.630 \ ; \quad \left(\frac{\mu}{\rho}\right)_{FeK_\alpha}^{specimen} = 8.148 \ (\chi = 12.676) \ m^2 kg^{-1}$$

$$\left(\frac{\mu}{\rho}\right)_{FeK^+}^{Ni} = 6.465 \ ; \quad \left(\frac{\mu}{\rho}\right)_{FeK^+}^{specimen} = 17.812 \ m^2 kg^{-1}$$

$$u_{Fe} = \frac{12.676}{17.812} = 0.7117 \ ; \quad u_{Ni} = 0.7348$$

Working out the Compositions

$$f\left(U_{FeK}^{0}\right) = 2.812 \ln 2.812 - 2.812 + 1 = 1.095; \quad f\left(U_{NiK}^{0}\right) = 0.7011$$

$$u_{Fe} U_{FeK}^{0} = 2.001 \qquad \frac{\ln\left(1 + u_{Fe} U_{FeK}^{0}\right)}{u_{Fe} U_{FeK}^{0}} = 0.5492$$

$$u_{Ni} U_{NiK}^{0} = 1.763 \qquad \frac{\ln\left(1 + u_{Ni} U_{NiK}^{0}\right)}{u_{Ni} U_{NiK}^{0}} = 0.5765$$

Equation 4.28 →

$$\left[\frac{44.288}{17.812}\left\langle 7.112 \times 1.095 \times 0.5492 - 8.333 \times 0.7011 \times 0.5765\right\rangle \right.$$

$$\left. + \frac{28.718}{32.961} \times 8.333 \times 0.7011 \times 0.5765\right] \Big/ 1.095$$

$$= 4.744$$

$$\frac{I_{FeK_\alpha}^{f}}{I_{FeK_\alpha}} = 4.34 \times 10^{-6} \times 0.8812 \times 55.85 \times 27.4 \times 4.744$$

$$= 0.028 \quad \text{in specimen}$$

NiK$_\alpha$ in Ni

$$\left(\frac{\mu}{\rho}\right)_{NiK_\alpha}^{Ni} = 5.652 \; (\chi = 8.793) \; m^2 kg^{-1}$$

$$\left(\frac{\mu}{\rho}\right)_{NiK^+}^{Ni} = 34.779 \; ; \; \left(\frac{\mu}{\rho}\right)_{NiK^-}^{Ni} = 4.195 \; m^2 kg^{-1}$$

$$\frac{r_{NiK} - 1}{r_{NiK}} = 0.8794 \; ; \; u = 0.2528 \; ; \; A_{Ni} = 58.69 \; ; \; E_K^{Ni} = 8.333 \; keV$$

$$\overset{0}{U}_{NiK} = 2.400 \; ; \; \frac{\ln\left(1 + u_{Ni}\overset{0}{U}_{NiK}\right)}{u_{Ni}\overset{0}{U}_{NiK}} = 0.6067$$

$$\frac{I_{NiK_\alpha}^f}{I_{NiK_\alpha}} = 4.34 \times 10^{-6} \times 0.8794 \times 58.69 \times 28 \times 8.333 \times \frac{34.779}{34.779} \times 0.6067$$

$$= 0.032 \quad \text{in Ni}$$

NiK_α in specimen

$$\left(\frac{\mu}{\rho}\right)_{NiK_\alpha}^{Fe} = 38.708 \; ; \; \left(\frac{\mu}{\rho}\right)_{NiK_\alpha}^{specimen} = 15.569 \; (\chi = 24.221) \; m^2 kg^{-1}$$

$$\left(\frac{\mu}{\rho}\right)_{NiK^+}^{Fe} = 28.718 \; ; \; \left(\frac{\mu}{\rho}\right)_{NiK^+}^{specimen} = 32.961 \; m^2 kg^{-1}$$

$$u_{Ni} = 0.7348 \; ; \; \frac{\ln\left(1 + u_{Ni}U_{NiK}^0\right)}{u_{Ni}U_{NiK}^0} = 0.5764 \; ; \; \overline{Z} = 27.4$$

$$\frac{I_{NiK_\alpha}^f}{I_{NiK_\alpha}} = 4.34 \times 10^{-6} \times 0.8794 \times 58.69 \times 27.4 \times 8.333 \times \frac{34.779}{32.961} \times 0.5764$$

$$= 0.031 \quad \text{in specimen}$$

Thus
$$F_{cd}(FeK\alpha) = \frac{1.030}{1.028} = 1.002$$

and
$$F_{cd}(NiK\alpha) = \frac{1.032}{1.031} = 1.001$$

Total ZAF correction (Equation (4.7))

$$F = F_s \times F_b \times F_a \times F_{ch} \times F_{co}$$

Fe K_α $F = 1.013 \times 1.008 \times 1.004 \times 0.859 \times 1.002 = 0.882$
Ni K_α $F = 0.996 \times 0.997 \times 1.043 \times 1.000 \times 1.001 = 1.037$

Thus the first corrected estimate of the true composition of the specimen is

$$C_{Fe}^m = 30.0 \text{ wt \%} \times 0.882 = 26.5 \text{ wt \% Fe}$$

$$C_{Ni}^m = 70.0 \text{ wt \%} \times 1.037 = 72.6 \text{ wt \% Ni}$$

Renormalising to 100 wt %, gives

$$C_{Fe}^m = 26.7 \text{ wt \%}$$

$$C_{Ni}^m = 73.3 \text{ wt \%}$$

I emphasise that there is no *need* to normalise: I chose to do so.

It is now necessary to iterate the calculation. With an efficient numerical analysis strategy (see, for example, Heinrich[48]) the solution will converge with sufficient accuracy after about three iterations. For example, the NBS COR2 computer programme, using a procedure and absorption coefficients rather different from those outlined above, converged after three iterations to the composition $C_{Fe}^m = 26.2$ wt%, $C_{Ni}^m = 71.9$ wt%, which, after renormalising, corresponds to $C_{Fe}^m = 26.7$ wt%, $C_{Ni}^m = 73.3$ wt%.

4.2.2.8 Final comments concerning chemical microanalysis of bulk specimens

The correction procedures I have implemented above are to some extent an arbitrary choice. I have tried to retain a straight line of logic starting from the physics of Chapter 2. Even within this restriction I could have selected differently - for example I could have used the full Bethe expression for the ionisation cross-section in the characteristic fluorescence correction, instead of the Green-Cosslett approximation, or I could have performed a full integration in the continuum fluorescence correction instead of the Springer approximation (as is done in the NBS COR 2 program, for example). My primary reason for making these two simplifications was to ease manual computation. I am certainly making no special claims for the particular procedure outlined above. What it is intended to impart is some feel for the *physics* underlying the various corrections. Without this, using a computer program to ZAF correct your microanalysis results is like walking along the edge of a precipice - the smallest mistake leads to disaster. If you would like to read about a fairly different approach to the ZAF corrections, try Love and Scott.[56] For an interesting comparison of various correction procedures, see Sevov *et al.*[57] A good recent book on quantitative bulk microanalysis is 'Electron probe quantitation', edited by Heinrich and Newbury,[58] where you will find, amongst other things, a description of current $\phi(\rho z)$ approaches to quantification.

In the worked example above both the X-ray peaks used were detected via WDX. A more usual way of proceeding nowadays is to use EDX for all elements with atomic numbers down to ~ that of sodium (11) and then to use WDX for the lighter elements only.

The difficult aspects of sections 4.2.2.1 - 4.2.2.5 are all to do with the way the electron's energy and position vary after its entry into the bulk specimen. Thus:

Working out the Compositions 171

Correction (z)	Stopping Power (S)	E(z)	I (z)	Path Length(z)
Stopping power	Bethe[5]			
Back-scattering		(empirical fit)		
Absorption			Lenard[18]	4 - (4-ϕ(0)) exp (-kρz) (Ref. 52)
Characteristic fluorescence		Whiddington[16]	Lenard[18]	
Continuum fluorescence		Whiddington[16]		

Table 4.2 Approximate descriptions of electron energy and position and how they enter the ZAF corrections

The two fundamental physical processes are the stopping power and elastic scattering. Stopping power is a function of s, path length, but E, I and path length itself are defined as functions of z, distance below the surface. In principle these last three could be derived from stopping power and elastic scattering, but this is not possible to do analytically. It can be done using Monte Carlo techniques, where each electron is tracked through the solid, and energy loss and elastic scattering are incorporated using random number generators (see Fig. 5.1). Monte Carlo calculations are very easy (and great fun) to perform but very difficult to interpret in any general way. (You can quickly confirm this for yourself by glancing at a few Monte Carlo papers.) The attraction of an analytical solution is that one can see relatively easily how varying each parameter will affect the final solution. An analytical solution is very powerful and contains a great deal of information. This is not true of the results of a Monte Carlo calculation. Their significance is largely restricted to the unique situation to which they have

Monte Carlo calculation? Partly for the reason I have just been explaining and partly because of the prohibitive computer time and expense.

Implicit in the electron's trajectory, of course, is the *spatial resolution* of the microanalysis. This will be dealt with (amongst other things) in the next and final chapter, Chapter 5.

Chapter 5 Some miscellaneous topics

5.1 Beam spreading

Scattering of the beam electrons by single atoms (see section 2.9) causes high angle deflections and is responsible for the spreading of the electron beam as it penetrates into the specimen. I was able to ignore this for the thin specimens used in transmission microscopes, because such spreading does not affect the corrections which were the subject of Chapter 4. This beam spreading in transmission specimens is, however, of interest in its own right. In the case of bulk specimens the single atom scattering is integral to the absorption corrections, including the absorption of fluorescing and fluoresced X-rays, and was taken into account in a fairly *ad hoc* manner, with a nod in the direction of the Rutherford formula (see section 4.2.2.3). Here too, however, beam spreading has its own intrinsic interest. In both transmission and bulk specimens beam spreading influences spatial resolution: the smallest area or volume of material which can be analysed. As with other corrections discussed earlier in this book, the transmission case is easier to attack and I will start with it.

5.1.1 Beam spreading in transmission specimens

Provided we are not worried about low angle scattering we can make the assumption that only one (if any) of the relatively rare high angle events occurs as the beam electron passes through the specimen. This is the *single scattering* approximation.[59] Referring back to section 2.9 and in particular to equation (2.18), we have that the cross-section per atom for scattering the beam electrons through an angle greater than θ is

$$\sigma(>\theta) = \frac{Z^2 e^4 (1-\beta^2)}{16\pi \, \varepsilon_0^2 m_0^2 c^4 \beta^4} \cot^2 \frac{\theta}{2}$$

174 Chemical Microanalysis Using Electron Beams

In a foil of thickness t, when there are N^v atoms/unit volume, the fraction of beam electrons thus scattered will be $p = N^v \sigma t$. A measure of beam spreading is the diameter of that circle on the exit surface of the specimen outside which a fraction p of the electrons emerge. A conventional choice for p is 10%, or 0.1.

If the scattering is taken to occur on average half-way through the foil then the beam broadening $b = t\theta$. Approximating $\cot^2 \theta/2$ as $4/\theta$ and combining the equations above,

$$b = \frac{Z e^2}{2 m_0 c^2 \varepsilon_0} \frac{\sqrt{1-\beta^2}}{\beta^2} \sqrt{\frac{N^v}{\pi p}} \, t^{3/2}$$

For beam energy $\ll 511$ keV,

$$\frac{\sqrt{1-\beta^2}}{\beta^2} = \frac{m_0 c^2}{2E} \qquad \text{(see equation (2.6))}$$

and then if $p = 0.1$,

$$b = 8 \times 10^{-12} \sqrt{N^v} \, \frac{Z}{E} \, t^{3/2} \quad \text{metres} \qquad (5.1)$$

where E is in keV, but otherwise the units are SI.

The Rutherford formula is conveniently manipulable and equation (5.1) is easily converted to other geometries etc. In fact, this is not generally worthwhile. Equation (5.1) should be taken as an order of magnitude indication only of where spatial resolution problems may arise and a useful indication of what the important controlling variables may be. Even then, vulgar reality may intrude - for example raising the beam voltage in a microscope will generally raise spurious microscope based X-rays and mask the improvement in b implied by equation (5.1). Remember, too, that the Rutherford equation does not apply at high beam voltages or at any voltage for heavy materials (see section 2.9). For light materials high angle single electron scattering will also contribute to beam spreading.

Example calculation 5.1

For a specimen of silicon 100 nm thick in a 100 kV TEM, what is the smallest probe size worth using? (The lattice parameter of silicon = 0.5431 nm.)

From equation (5.1):

$$b = 8 \times 10^{-12} \sqrt{N^v \frac{Z}{E}} t^{3/2} \text{ metres}$$

(Here p (see above) is 1%.)

$$N^v = 8/(0.5431 \times 10^{-9})^3 = 4.994 \times 10^{28} \text{ atom m}^{-3}$$

$$b = 8 \times 10^{-12} \sqrt{4.994 \times 10^{28} \frac{14}{100}} (100 \times 10^{-9})^{3/2}$$

$$\sim 8 \text{ nm}$$

Evidently a beam size much less than this would be rather pointless.

5.1.2 Beam spreading in bulk specimens

This is more complicated than for transmission specimens. The beam spreading is much greater and therefore the spatial resolution of SEM microanalysis is inherently much worse than that of TEM microanalysis, regardless of the initial size of the beam. Beam spreading in bulk specimens is controlled by competition between elastic scattering by single atoms and the slowing down of the beam electrons by inelastic electron interactions. The simplest way to calculate this is via a Monte Carlo simulation. An example is shown in Fig. 5.1.

A simple estimate of beam spreading is frequently all that is required and based on Lenard's law (see equation 2.10) Reed[18] has given the formula:

176 *Chemical Microanalysis Using Electron Beams*

$$b = \frac{2 \times 10^{-7}}{\rho} \left(E_0^{15} - E_{nl}^{15} \right) \quad \text{metres} \quad (5.2)$$

where E_o and E_{nl} are in keV and ρ is in Mg m^{-3}. Equation (5.2) corresponds to a probability, p (see section 5.1.1) of 1%. E_{nl} is included because after E drops below E_{nl} the spatial resolution of X-ray microanalysis is unaffected by the electron's trajectory. If X-ray fluorescence occurs equation (5.2) may be a substantial underestimate.

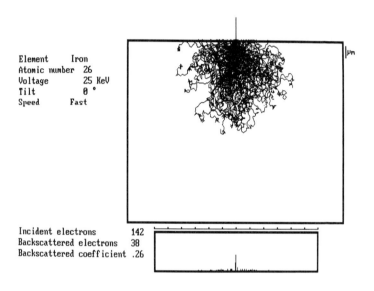

Fig. 5.1 Monte Carlo simulated electron beam trajectories in a bulk specimen of iron. Note that the trajectories of the electrons emerging from the surface of the specimen have been truncated soon after their exit. (Output from Institute of Materials software series disc, reproduced by kind permission of the author, Professor F.J. Humphreys.)

5.2 Low energy X-rays

Everything is more difficult with low energy (soft) X-rays:

- they are very strongly absorbed in the specimen: any surface

Some Miscellaneous Topics 177

- layers due to electropolishing, for example, or corrosion will assume a misleading importance.
- the large absorption coefficients are often not known very accurately.
- for the same reason any surface roughness will be especially deleterious.
- EDX detector windows will absorb strongly.
- EDX counting electronics will struggle with the small pulses produced.
- even WDX spectrometers produce small, relatively wide peaks with poor peak-to-background ratios.

Soft characteristic X-rays are always emitted, but for medium and high atomic numbers there are always alternative higher energy ones available. For low atomic number elements, however (say $Z < 10$), the highest energy (K) X-rays are of low energy and we must confront the rather daunting list of problems above. Fig. 5.2 shows a WDX spectrometer nitrogen K peak.

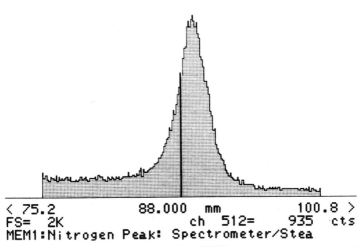

Fig. 5.2 A nitrogen K peak measured using a WDX spectrometer (compare with Fig. 3.9(a)). The diffracting crystal was lead stearate. An improved peak height could have been obtained using an artificially constructed layer crystal. (Spectrum courtesy M.G. Hall.)

178 *Chemical Microanalysis Using Electron Beams*

The poorer peak-to-background ratio as compared with the spectrum shown in Fig. 3.9(a) is caused by absorption in the detecting crystal and broadening of the angular acceptance range by defects in it. This spectrum was detected using a lead stearate crystal. Artificial multilayer crystals are now available which improve substantially soft X-ray peak heights.

For soft X-ray analysis EDX detectors may be fitted with very thin windows (see p. 76-77) or the window may even be removed altogether. In this way elements with Z as low as 4 (Be) may be analysed. No window at all may render the detecting crystal performance open to degradation by light from the filament or the specimen and by surface contamination. Fig. 5.3 shows an EDX spectrum from a specimen of alumina supported on a carbon film. It is also certainly worth optimising the electronics of your EDX system for soft X-ray detection.

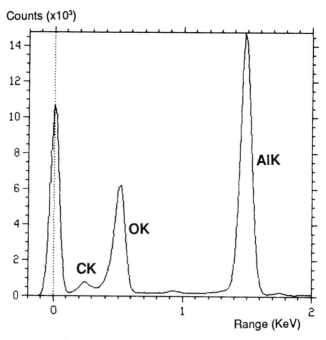

Fig. 5.3 A thin window EDX spectrum from a particle of alumina supported on a thin carbon film. In this case the thin window was a 100 nm film of aluminium supported on a nickel grid.

5.3 Alchemi

This is the name devised by its originators[60] for a method of locating atoms within unit cells by exploiting the electron channelling which results from Bragg reflection. You may recall (section 2.10 and Fig. 2.15) that setting the crystal with the deviation parameter 's' positive results in the majority of the electrons being channelled between the reflecting planes of atoms. A -ve s orientation channels the electrons along the planes. Because X-ray production, involving as it does the ejection of core electrons, is localised close to the atom nuclei, it will be enhanced for negative s. If different atomic species are present at various positions in the unit cell, tilting the crystal will change the relative incident electron intensities each species sees. It is frequently possible to extract useful information about atom positions without the necessity to calculate accurately what the distribution of beam electrons across the unit cell actually is.

For example, the data in Table 5.1 show how the Al, Ti and Nb X-ray intensities vary with orientation about a superlattice Bragg position for a specimen of TiAl in which has been dissolved some Nb.[61] The fact that the Ti/Nb ratio remains constant with changes in orientation, whereas the Ti/Al ratio does not, shows that the Nb inhabits the Ti rather than the Al sublattice in this ordered intermetallic compound. If a solute is divided between two sublattices a simple extension of this logic enables the partition ratio to be determined. Note, however, that if the monitored X-rays have very different energies the localisations of the ionisation events to the nuclei will be different and this of itself can give an effect on tilting the crystal, although this either does not happen here or it cancels out with a small mixing effect.

Reflecting planes	Sign. of s	AlK/TiK	NbK/TiK
{110}	+ve	0.432	0.068
	-ve	0.264	0.066
(001)	+ve	0.447	0.067
	-ve	0.339	0.065

Table 5.1 The ratios of Al : Ti : Nb X-ray intensities as a crystal of Ti-52Al-5Nb is tilted about two superlattice reflection Bragg positions.[61]

180 *Chemical Microanalysis Using Electron Beams*

5.4 Electron energy loss spectroscopy (EELS)[11,40]

The interactions between the fast beam electrons and either individual electrons or the plasma of conduction or valence electrons in the solid, which were described in sections 2.2 - 2.4, result in the beam electrons losing substantial amounts of energy. These may be measured using an electron energy spectrometer at the end of a transmission microscope. Such a spectrometer is visible in Figs. 3.2(a) and 3.3. A schematic diagram of a spectrometer is shown in Fig. 5.4.

Spectrometers may be *serial*, where the spectrum is scanned over the detector, or *parallel*, where an array of detectors measures the whole spectrum at once (much better). An example of a spectrum is shown in Fig. 5.5(a). At the low energy loss end, following the zero loss peak,* appear

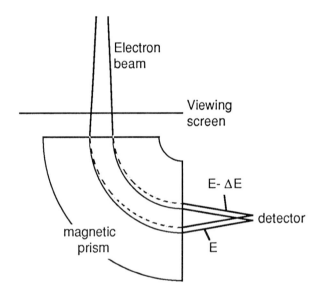

Fig. 5.4 A schematic diagram of an electron energy loss spectrometer (EELS). For serial EELS, the spectrum is scanned across a single detector. For parallel EELS (PEELS) there is an array of detectors. (Adapted from Egerton.[11])

---

*The width of this is due to the energy spread of the incident beam and to the finite energy resolution of the spectrometer.

the plasmon losses. These are peaks, just like in an EDX spectrum. A magnified and rather clearer example of plasmon losses appeared in Fig. 2.6. The probability of n plasmons being excited in a foil of thickness t is

$$P_n = \frac{1}{n!} \left(\frac{t}{\Lambda}\right)^n \exp\left(-\frac{t}{\Lambda}\right) \tag{5.3}$$

where Λ is the mean free path for plasmon creation. If Λ is known, t can be calculated from the size of the plasmon peaks (see Example calculation 5.2 below).

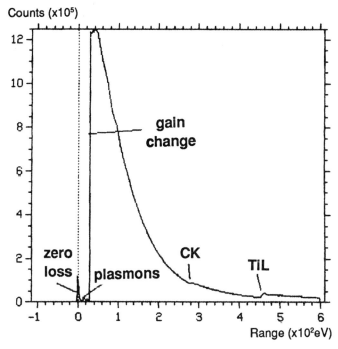

Fig. 5.5(a) An EEL spectrum from TiC showing zero energy loss peak, plasmon peaks and inner shell ionisation edges. Note the gain change between the plasmon peaks and the inner shell edges. (Spectrum courtesy A.J. Burbery and M.H. Loretto.)

182 *Chemical Microanalysis Using Electron Beams*

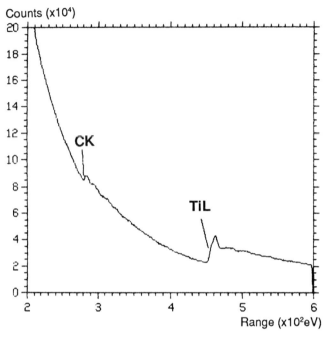

Fig. 5.5(b) An enlarged view of part of Fig. 5.5(a) showing the two ionisation edges.

At higher energy losses appears a series of *edges* corresponding to the ionisation of various energy levels in the specimen's atoms (Fig 5.5(b)). These are edges rather than peaks because the beam electrons may expend any energy above the minimum defined by the edge in ionising the atom. By measuring the number of electrons (the area) under the edge and comparing it with that forming the zero loss peak, the total mass of the element within the irradiated volume of the specimen may be deduced. This information is rather more detailed than that normally derived from an EDX spectrum: clearly by taking the ratios of masses of elements, relative concentrations may also be obtained via EELS.

To measure and interpret the size of the edge the background has to be removed. The background is made up of ionisation events belonging to a multitude of lower energy edges, which are usually individually too minor to be noticeable. This background cannot be predicted accurately but is

Some Miscellaneous Topics

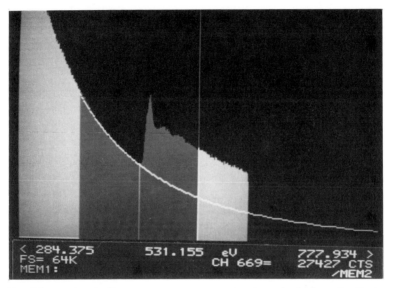

Fig. 5.5(c) A fitted AE^{-r} background extrapolated under an ionisation edge. In the current example $r \sim 3$. This is common.

found to have the form AE^{-r}. 'A' and 'r' are determined via a logarithmic plot on the high energy (low energy loss) side of the peak. The background is then extrapolated under the edge of interest. This is illustrated in Fig. 5.5(c), which shows the fitted background extrapolated under an edge. The area under the edge is measured over a limited energy range - clearly the extrapolation becomes more uncertain as the energy interval below the edge increases. ~ 100 eV is normal. Following subtraction of the background the measured area (Fig. 5.5(d)) is divided by the zero loss intensity and the ratio converted to number of atoms via a *partial* cross-section. A partial cross-section is appropriate because:

(i) a finite energy loss range has been used and
(ii) the EELS detector intercepts a finite solid angle - some of the ionising electrons pass outside the detector.

184 *Chemical Microanalysis Using Electron Beams*

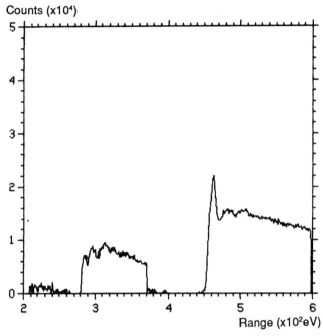

Fig. 5.5(d) The stripped CK and TiL edges.

Expressions for the partial ionisation cross-section of varying sophistication may be used. A simple approach is to multiply the total cross-section for ionisation (equations (2.4) or (2.5)) by energy and angle efficiency factors:[11]

If the shape of the edge after background subtraction is fitted to BE^{-s}, then clearly:

$$\text{Energy } \Delta E : \eta(\Delta E) = 1 - \left(1 + \frac{\Delta E}{E_{nl}}\right)^{1-s} \quad (5.4a)$$

Also

$$\text{Angle } \phi : \eta(\phi) = \frac{\ln\left(1 + \left(\frac{\phi}{\theta_E}\right)^2\right)}{\ln\left(\frac{2}{\theta_E}\right)} \quad (5.4b)$$

Some Miscellaneous Topics 185

where $\quad \theta_E = \dfrac{E_{nl} + \dfrac{\Delta E}{2}}{hk} \quad$ (k = wave vector of incident electrons)

Alternatively, the *hydrogenic approximation,* introduced by Bethe[5] in his original paper, may be applied. In this approximation the ionised atom and the removed electron are considered as forming a hydrogen atom with nuclear charge Z corrected by the screening effect of the remaining (Z-1) electrons. This approximation has been implemented as simple computer program (see Appendix A of Egerton's book[11] where other useful EELS processing programs, beyond the scope of this section, are listed).

By way of comparison, the equivalent EDX spectrum is shown in Fig. 5.6.

The spectrum shown in Fig. 5.5 is analysed in Example calculation 5.3 below.

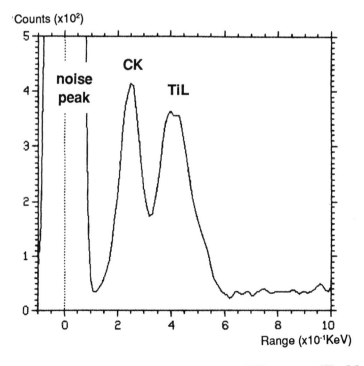

Fig. 5.6 An EDX spectrum from TiC corresponding to the EEL spectrum of Fig. 5.5

Example calculation 5.2

What was the specimen thickness for the EEL spectrum shown in Fig. 2.6? The beam voltage was 400 kV.

Recall equation (5.3):

$$P_n = \frac{1}{n!}\left(\frac{t}{\Lambda}\right)^n \exp\left(-\frac{t}{\Lambda}\right) \qquad (5.3)$$

Then
$$\frac{P_1}{P_0} = \frac{t}{\Lambda}$$

Normally one would measure Λ for a given material using plasmon peaks and an independent measurement. Here I shall estimate Λ (see Egerton[11]):

$$\frac{1}{\Lambda} = \frac{E_p}{m_0 v^2 a_H} \ln\left(\frac{\theta_c}{\theta_E}\right)$$

where E_p = plasmon energy = 15.2 eV here (see Fig. 2.6)

θ_c = cut-off angle = K/K_c and $K_c = (N^v)^{1/3}$

and N_v = number of electrons / unit volume

$$= \frac{m\varepsilon_0}{e^2}\left(\frac{2\pi E_p}{h}\right)^2$$

$$\theta_E = \frac{E_p}{m v^2}$$

and a_H = Bohr radius = 52.9 pm (Appendix A, Table A.1)

Then n = 1.675×10^{29} electrons m^{-3}

Some Miscellaneous Topics 187

$$K_c = 5.512 \times 10^9 \text{ m}^{-1}$$
$$K = 6.083 \times 10^{11} \text{ m}^{-1}$$
$$\theta_c = 9.061 \times 10^{-3} \text{ rads}$$
$$\theta_E = 2.434 \times 10^{-5} \text{ rads}$$
$$\beta = 0.8279$$
$$v = 2.482 \times 10^8 \text{ m s}^{-1}$$
and $\Lambda = 206$ nm

From Fig. 2.6 $\quad \dfrac{P_1}{P_0} = 0.61 = \dfrac{t}{\Lambda}$

and $\quad t = 126$ nm

Note that the other ratios P_2/P_0 and P_3/P_0 in Fig. 2.6 are consistent with equation (5.3).

Example calculation 5.3

Estimate the composition of the TiC specimen whose EEL spectrum is shown in Fig. 5.5.
(The carbon K edge contained 602814 electrons between the edge at 284 keV and (284+78) eV. The titanium L edge contained 1128070 electrons between the edge at 455 eV and (455+73) eV. The zero loss peak contained 1.67×10^8 electrons and the spectrometer acceptance angle was 13 mrads.)

The efficiency factors referred to in equations (5.4) are really only applicable to K edges. Here I choose to use the SIGMAK2 and SIGMAL2 programs for partial ionisation cross-sections reproduced in Appendix A of Egerton.[11] Then :

$$\sigma_{CK} = (78 \text{ eV, } 13 \text{ mrad}) = 4.93 \times 10^{-25} \text{ m}^2$$

$$\sigma_{TiL} = (73 \text{ eV, } 13 \text{ mrad}) = 5.58 \times 10^{-25} \text{ m}^2$$

Then number of C atoms m^{-2} = $\dfrac{602814}{1.67 \times 10^8 \times 4.93 \times 10^{-25}}$ = 7.3×10^{21}

and number of Ti atoms m^{-2} = $\dfrac{1128070}{1.67 \times 10^8 \times 5.58 \times 10^{-25}}$ = 1.2×10^{22}

This implies a composition of $TiC_{0.6}$.

5.5 Some valedictory comments

It might be appropriate at this stage to return to Chapter 1 and re-read the comparison of TEM and SEM microanalysis, which was included there for completeness' sake but the reasons for which you should now appreciate better.

If there is one piece of advice I would urge on you, it is to be always very sceptical of your results and not to accept blindly the results of computer processing by proprietary programs whose details you will not have access to. Computer output has a spurious and seductive respectability all of its own. A few back-of-envelope calculations using the expressions and data described in this book (and others) may save you from wasting hours analysing further meaningless or invalid results.

Not wishing to end on a negative note, it is equally true that a thorough familiarity with your analytical equipment and the relatively simple physics underlying its exploitation will enable you to form an accurate judgment of the reliability and significance of your chemical microanalysis.

References

1. E. Rutherford: *Phil. Mag.*, 1911, **21**, 669-688.

2. R.D. Evans: 'The atomic nucleus', 1955, New York, McGraw-Hill.

3. N. Bohr: *Phil. Mag.*, 1913, **25**, 10-31.

4. A selection of quantum mechanics books:

   R.H. Bransden and C.J. Joachim: 'Introduction to Quantum Mechanics', 1989, Harlow, Longman.

   H. Clark: 'A First Course in Quantum Mechanics', 1982, Wokingham, Van Nostrand Reinhold.

   J.M. Cassels: 'Basic Quantum Mechanics', 1982, London, Macmillan.

   R.H. Dicke and J.P. Wittke: 'Introduction to Quantum Mechanics', 1974, Reading (Mass), Addison.

   A.P. French and E.F. Taylor: 'An Introduction to Quantum Physics', 1978, Wokingham, Van Nostrand.

   H. Haken and H.C. Wolf: 'Atomic and Quantum Physics', 2 edn, 1987, New York, Springer.

   (A useful criterion in choosing a book on quantum mechanics is to pick the one with the most pictures.)

5. H.A. Bethe: *Ann. Phys.*, 1930, **5**, 325-400 (available in English translation from British Lending Library).

6. M. Inokuti: *Rev. Mod. Phys.*, 1971, **43**, 297-347.

   M. Inokuti, Y. Itikawa and J.E. Turner: *Rev. Mod. Phys.*, 1978, **50**, 23-26.

7. M. Inokuti and S.T. Manson: in 'Electron Beam Interactions with Solids for Microscopy, Microanalysis and Microlithography', Proc. 1st Pfefferkorn Conf., Monterey, California, 1982, (ed. D.F. Kyser *et al.*), 1-17, 1984, AMF O'Hare, Illinois, SEM Inc.

8. J.H. Paterson, J.N. Chapman, W.A.P. Nicholson and J.M. Titchmarsh: *J. Microsc.*, 1989, **154**, 1-17.

9. M. Green and V.E. Cosslett: Proc. Phys. Soc., 1961, **78**, 1206-1214.

10. C.J. Powell: *Rev. Mod. Phys.*, 1976, **48**, 33-47.

11. R.F. Egerton: 'Electron Energy-Loss Spectroscopy', 1986, New York, Plenum.

12. H.A. Bethe: *Z. Physik*, 1932, **76**, 293-299.

13. N.J. Zaluzec: in 'Analytical Electron Microscopy 1984', (ed. D.B. Williams and D.C. Joy), 279-284, 1984, San Francisco, San Francisco Press.

14. F. Bloch: *Z. Physik*, 1933, **81**, 363-376.

15. H.A. Bethe and J. Ashkin: 'Experimental Nuclear Physics', 1953, New York, Wiley.

16. R. Whiddington: *Proc. Royal Soc.*, 1912, **A86**, 360-370.

17. C.J. Humphreys: Unpublished work.

18. S.J.B. Reed: 'Electron Microprobe Analysis', 2 edn, 1992, Cambridge, Cambridge University Press.

19. H.W. White: 'Introduction to Atomic Spectra'; 1934, New York, McGraw-Hill.

20. N.A. Dyson: 'X-rays in Atomic and Nuclear Physics', 2 edn, 1990, Cambridge, Cambridge University Press.

21. W. Bambynek, B. Crasemann, R.W. Fink, H.-U. Freund, H. Mark, C.D. Swift, R.E. Price and P. Venugopala Rao: *Rev. Mod. Phys.*, 1972, **44**, 716-813.

22. M.O. Krause: *J. Phys. Chem. Ref. Data*, 1979, **8**, 307-327.

23. T.P. Schreiber and A.M. Wims: *X-ray Spectrometry*, 1982, **11**, 42-45.

24. T.P. Thinh and J. Leroux: *X-ray Spectrometry*, 1979, **8**, 85-91.

25. K.F.J. Heinrich: in 'Proc. 11th Int. Cong. on X-ray Optics and Microanalysis', (ed. J.D. Brown and R.H. Packwood), 1987, 67-119, London (Ontario).

26. B.L. Henke, P. Lee, T.J. Tanaka, R.L. Shimabukuro and B.K. Fujikawa: *Atomic Data and Nuclear Data Tables*, 1982, **27**, 1-144.

27. H.A. Kramers: *Phil. Mag.*, 1923, **46**, 836-871.

28. A. Sommerfeld: *Ann. Phys.*, 1931, **11**, 257-330.

29. H.A. Bethe and W. Heitler: *Proc. Royal Soc.*, 1934, **A146**, 83-112.

30. J.C. Russ: 'Fundamentals of Energy Dispersive X-ray Analysis', 1984, London, Butterworths.

31. N.F. Mott: *Proc. Royal. Soc.*, 1932, **A135**, 429-458.

32. W.A. McKinley, Jr. and H. Feshbach: *Phys. Rev.*, 1948, **74**, 1759-1763.

33. N.F. Mott and H.S.W. Massey: 'The Theory of Atomic Collisions Vols. 1 and 2', 3 edn, 1965, Oxford, Oxford University Press.

34. P. Rez: in 'Electron Beam Interactions with Solids for Microscopy, Microanalysis and Microlithography' (Proceedings of the 1st Pfefferkorn Conference, Monterey, California, 1982), (ed. D.F. Kyser *et al.*), 43-49, 1984, AMF O'Hare, Illinois, SEM Inc.

See also C.R. Bradley: 'Calculations of Atomic Sputtering and Displacement Cross-sections in Solid Elements by Electrons with Energies from Threshold to 1.5MV', Argonne National Laboratory Report ANL-88-48, 1988, Argonne, Illinois, U. S. Department of Energy.

35. J.A. Doggett and L.V. Spencer: *Phys. Rev.*, 1956, **103**, 1597-1601.

36. F. Lenz: *Z. Naturf.*, 1954, **9A**, 185-204.

37. P.B. Hirsch, A. Howie, R.B. Nicholson, D. Pashley and M.J. Whelan: 'Electron Microscopy of Thin Crystals', 1977, Huntington (NY), Krieger.

38. S.J.B. Reed and N.G. Ware: *J. Phys. E.*, 1972, **5**, 582-584.

39. Some good general references on electron beam microanalysis:

    J.P. Eberhart: 'Structural and Chemical Analysis of Materials', 1991, Chichester, Wiley.

    P.J. Goodhew and F.J. Humphreys: 'Electron Microscopy and Analysis', 2 edn, 1988, London, Taylor & Francis.

    J.J. Hren, J.I. Goldstein and D.C. Joy (editors): 'Introduction to Analytical Electron Microscopy', 1979, New York, Plenum.

    M.H. Loretto: 'Electron Beam Analysis of Materials', 1984, London, Chapman and Hall.

40. D.J. Williams: 'Practical Analytical Electron Microscopy in Materials Science', 1987, Herndon (VA), Tech Books.

41. Y. Li and M.H. Loretto: (private communication)
    Y. Li: M. Phil. Thesis, 1991, University of Birmingham.

42. N.J. Zaluzec: *EMSA Bulletin*, 1985, **15**, 67-83.

43. G. Cliff and G.W. Lorimer: *J. Microsc.*, 1975, **103**, 203-207.

44. P.M. Kelly, A. Jostsons, R.G. Blake and J.G. Napier: *Phys. Stat. Sol.* (a), 1975, **31**, 771-780.

45. M.E. Twigg: in 'Analytical Electron Microscopy 1984', (ed. D.B. Williams and D.C. Joy), 1984, San Francisco, San Francisco Press, 325-326.

46. C. Nockolds, M.J. Nasir, G. Cliff and G.W. Lorimer: in 'EMAG'79, Inst. Phys. Conf. Series No. 52', (ed. T. Mulvey), 1980, 417-420, IOP, Bristol.

47. C.E. Fiori, R.L. Myklebust and K.F.J. Heinrich: *Anal. Chem.*, 1976, **48**, 172-176.

48. K.F.J. Heinrich: 'Electron Beam X-ray Microanalysis', 1981, New York, Van Nostrand Reinhold.

49. R.R. Wilson: *Phys. Rev.*, 1941, **60**, 749-753.

50. M.J. Berger and S.M. Seltzer: in 'Studies in Penetration of Charged Particles in Matter', National Research Council Publication 1133, 205-268, 1964, Washington, National Academy of Sciences.

51. P. Duncumb and S.J.B. Reed: in 'Quantitative Electron Probe Microanalysis' (ed. K.F.J. Heinrich), NBS Spec Pub. 298, 1968, Washington, US Dept. of Commerce, 133-154.

52. J. Philibert: in 'X-ray Optics and X-ray Microanalysis', (ed. H.H. Pattee, V.E. Cosslett and A. Engström), 1963, New York, Academic Press, 379-392.

53. K.F.J. Heinrich: NBS Technical Note 521, 1970, Washington DC, NBS (US Dept of Commerce).

54. S.J.B. Reed and J.V.P. Long: in 'X-ray Optics and X-ray Microanalysis', (ed. H.H. Pattee, V.E. Cosslett and A. Engström), 1963, New York, Academic Press, 317-327.

55. G. Springer: *Neues Jahrb. Mineral., Abhandl.*, 1967, **106**, 241-256.

and : in Proc. 6th Int. Conf. on X-ray Optics and Microanalysis, (ed. G. Shinoda, K. Murata and R. Shimizu), 1972, Tokyo, Univ. Press, 141-146.

56. G. Love and V.D. Scott: *J. Phys. D.*, 1978, **11**, 1369-1376.

57. S. Sevov, H.P. Degischer, H -J. August and J. Wernisch: *Scanning*, 1989, **11**, 123-134.

58. K.F.J. Heinrich and D.E. Newbury (eds): 'Electron Probe Quantitation', 1991, New York, Plenum.

59. J.I. Goldstein, J.L. Costley, G.W. Lorimer and S.J.B. Reed: *Scanning Electron Microscopy*, 1977, **1**, 315-324.

60. J.C.H. Spence and J. Taftø: *J. Microsc.*, 1983, **130**, 147-154.

61. D.G. Konitzer, I.P. Jones and H.L. Fraser: *Scripta Met.*, 1986, **20**, 265-268.

62. G.G. Johnson, Jr and E.W. White: 'X-ray Emission Wavelength and KeV Tables for Non-diffractive Analysis', ASTM Data Series DS46, 1970, Mars (Pa), ASTM.

63. M. Abramowitz and I.A. Stegun (eds): 'Handbook of Mathematical Functions', 1965, New York, Dover.

Appendix A

Useful Data

Table A.1   Some useful constants and formulae

Table A.2   Atomic numbers and relative atomic masses

Table A.3   Characteristic X-ray energies

Table A.4   Fluorescence yields

Table A.5   Partition factors

Table A.6   Mass absorption coefficients and ionisation energies for X-rays

Table A.7   The logarithmic integral

Electronic charge (i.e. 1eV)	e	= =	1.6021×10^{-19} C 1.6021×10^{-19} J
Rest mass of electron	m_0	=	9.109×10^{-31} kg
Permittivity of free space (C^2 J^{-1} m^{-1})	ε_0	=	8.854×10^{-12} F m^{-1}
Velocity of light	c	=	2.998×10^8 m s^{-1}
Planck's constant	h	=	6.626×10^{-34} J s
Avogadro's constant	N_0	=	6.022×10^{23} mol^{-1}

and implicit in the above:

Rest energy of electron $= m_0 c^2 = 511$ keV

Fine structure constant $= e^2/(2\varepsilon_0 hc) \sim 1/137$ (often referred to as α, but not in this book)

Bohr radius $a_H = \dfrac{4\pi\varepsilon_0 \left(\dfrac{h}{2\pi}\right)^2}{m_0 e^2} = 52.9$ pm

Electron wavelength
$$\lambda = \dfrac{1}{\left\{2 m_0 E \left(1 + \dfrac{E}{2 m_0 c^2}\right)\right\}^{1/2}}$$

$$= \dfrac{3.878 \times 10^{-11}}{\sqrt{E}\sqrt{1 + 0.9785 \times 10^{-3} E}} \text{ metres}$$

(E in keV)

Table A.1 Some useful constants and formulae.

A.2 Atomic Numbers and Relative Atomic Masses

Element	Atomic Number Z	Relative Atomic Mass	Element	Atomic Number Z	Relative Atomic Mass	Element	Atomic Number Z	Relative Atomic Mass
H	1	1.01	Ge	32	72.61	Eu	63	151.97
He	2	4.00	As	33	74.92	Gd	64	157.25
Li	3	6.94	Se	34	78.96	Tb	65	158.93
Be	4	9.01	Br	35	79.90	Dy	66	162.50
B	5	10.81	Kr	36	83.80	Ho	67	164.93
C	6	12.01	Rb	37	85.47	Er	68	167.26
N	7	14.01	Sr	38	87.62	Tm	69	168.93
O	8	16.00	Y	39	88.91	Yb	70	173.04
F	9	19.00	Zr	40	91.22	Lu	71	174.97
Ne	10	20.18	Nb	41	92.91	Hf	72	178.49
Na	11	22.99	Mo	42	95.94	Ta	73	180.95
Mg	12	24.31	Tc	43	(98)	W	74	183.85
Al	13	26.98	Ru	44	101.07	Re	75	186.21
Si	14	28.09	Rh	45	102.91	Os	76	190.2
P	15	30.97	Pd	46	106.42	Ir	77	192.22
S	16	32.07	Ag	47	107.87	Pt	78	195.08
Cl	17	35.45	Cd	48	112.41	Au	79	196.97
Ar	18	39.95	In	49	114.82	Hg	80	200.59
K	19	39.10	Sn	50	118.71	Tl	81	204.38
Ca	20	40.08	Sb	51	121.75	Pb	82	207.2
Sc	21	44.96	Te	52	127.60	Bi	83	208.98
Ti	22	47.88	I	53	126.90	Po	84	(209)
V	23	50.94	Xe	54	131.29	At	85	(210)
Cr	24	52.00	Cs	55	132.91	Rn	86	(222)
Mn	25	54.93	Ba	56	137.33	Fr	87	(223)
Fe	26	55.85	La	57	138.91	Ra	88	226.03
Co	27	58.93	Ce	58	140.12	Ac	89	227.03
Ni	28	58.69	Pr	59	140.91	Th	90	232.04
Cu	29	63.55	Nd	60	144.24	Pa	91	231.04
Zn	30	65.39	Pm	61	(145)	U	92	238.03
Ga	31	69.72	Sm	62	150.36			

Table A 2 Atomic numbers and relative atomic masses of the elements. Numbers in parentheses refer to most stable isotope.

Z	El	M			L			K			El	Z
		M_α	M_β	M_γ	L_α	$L_{\beta 1}$	$L_{\beta 2}$	K_α	K_α	$K_{\beta 1}$		
1	H										H	1
2	He										He	2
3	Li							0.052			Li	3
4	Be							0.110			Be	4
5	B							0.185			B	5
6	C							0.277			C	6
7	N							0.392			N	7
8	O							0.525			O	8
9	F							0.677			F	9
10	Ne							0.848			Ne	10
11	Na							1.041		(1.067)	Na	11
12	Mg							1.253		(1.295)	Mg	12
13	Al							1.486		(1.553)	Al	13
14	Si							1.739		(1.829)	Si	14
15	P							2.013		(2.136)	P	15
16	S							2.307		2.464	S	16
17	Cl							2.621		2.815	Cl	17
18	Ar							2.957		3.190	Ar	18
19	K							3.312		3.589	K	19
20	Ca				0.341	0.345		3.687	3.691	4.012	Ca	20

A.3 Characteristic X-ray Energies

Z	El	M			L			K			El	Z
		M_α	M_β	M_γ	L_α	$L_{\beta 1}$	$L_{\beta 2}$	K_α	$K_{\alpha'}$	$K_{\beta 1}$		
21	Sc				0.395	0.400		4.085	4.090	4.460	Sc	21
22	Ti				0.452	0.458		4.504	4.510	4.931	Ti	22
23	V				0.511	0.519		4.944	4.951	5.426	V	23
24	Cr				0.573	0.583		5.405	5.414	5.946	Cr	24
25	Mn				0.637	0.649		5.887	5.898	6.489	Mn	25
26	Fe				0.705	0.718		6.390	6.403	7.057	Fe	26
27	Co				0.776	0.791		6.914	6.929	7.648	Co	27
28	Ni				0.851	0.869		7.460	7.477	8.263	Ni	28
29	Cu				0.930	0.950		8.026	8.046	8.904	Cu	29
30	Zn				1.012	1.034		8.614	8.637	9.570	Zn	30
31	Ga				1.098	1.125		9.223	9.250	10.263	Ga	31
32	Ge				1.188	1.218		9.854	9.885	10.980	Ge	32
33	As				1.282	1.317		10.506	10.542	11.724	As	33
34	Se				1.379	1.419		11.179	11.220	12.494	Se	34
35	Br				1.480	1.526		11.876	11.922	13.289	Br	35
36	Kr				1.586	1.636		12.596	12.648	14.110	Kr	36
37	Rb				1.694	1.752		13.333	12.393	14.959	Rb	37
38	Sr				1.806	1.871		14.095	14.163	15.833	Sr	38
39	Y				1.922	1.995		14.880	14.956	16.735	Y	39
40	Zr				2.042	2.124	(2.219)	15.688	15.772	17.665	Zr	40

Z	El	M			L			K			El	Z
		M_α	M_β	M_γ	L_α	$L_{\beta1}$	$L_{\beta2}$	$K_{\alpha2}$	K_α	$K_{\beta1}$		
41	Nb			0.355	2.166	2.257	(2.367)	16.518	16.612	18.619	Nb	41
42	Mo				2.293	2.394	(2.518)	17.371	17.476	19.605	Mo	42
43	Tc				2.424	2.536	(2.674)	18.248	18.364	20.615	Tc	43
44	Ru			0.461	2.558	2.683	(2.835)	19.147	19.276	21.653	Ru	44
45	Rh			0.496	2.696	2.834	3.001	20.070	20.213	22.720	Rh	45
46	Pd			0.532	2.838	2.990	3.171	21.017	21.174	23.815	Pd	46
47	Ag			0.568	2.984	3.150	3.347	21.987	22.159	24.938	Ag	47
48	Cd			0.606	3.133	3.316	3.528	22.980	23.170	26.091	Cd	48
49	In				3.286	3.487	3.713	23.998	24.206	27.271	In	49
50	Sn			0.691	3.443	3.662	3.904	25.040	25.267	28.481	Sn	50
51	Sb			0.733	3.604	3.843	4.100	26.106	26.355	29.721	Sb	51
52	Te			0.778	3.769	4.029	4.301	27.197	27.468	30.990	Te	52
53	I				3.937	4.220	4.507	28.312	28.607	32.289	I	53
54	Xe				4.109	4.422	4.720	29.453	29.774	33.619	Xe	54
55	Cs				4.286	4.619	4.935	30.620	30.968	34.981	Cs	55
56	Ba			0.972	4.465	4.827	5.156	31.812	32.188	36.372	Ba	56
57	La	0.833	0.854	(1.026)	4.650	5.041	5.383	33.028	33.436	37.795	La	57
58	Ce	0.883	0.902	(1.075)	4.839	5.261	5.612	34.273	34.714	39.251	Ce	58
59	Pr	0.929	0.949	(1.127)	5.033	5.488	5.849	35.544	36.020	40.741	Pr	59
60	Nd	0.978	0.995	(1.180)	5.229	5.721	6.088	36.841	37.355	42.264	Nd	60

A.3 Characteristic X-ray Energies 201

Z	El	M			L			K		
		M_α	M_β	M_γ	L_α	$L_{\beta 1}$	$L_{\beta 2}$	K_α	$K_{\alpha'}$	$K_{\beta 1}$
61	Pm	1.032	1.055		5.432	5.960	6.338	38.165	38.718	43.818
62	Sm	1.081	1.100	(1.291)	5.635	6.204	6.586	39.516	40.111	45.405
63	Eu	1.131	1.153	(1.346)	5.845	6.455	6.842	40.895	41.535	47.030
64	Gd	1.185	1.209	(1.402)	6.056	6.712	7.102	42.302	42.989	48.688
65	Tb	1.240	1.266	(1.461)	6.272	6.977	7.365	43.737	44.474	50.374
66	Dy	1.293	1.325	(1.522)	6.494	7.246	7.634	45.200	45.991	52.110
67	Ho	1.347	1.383	(1.576)	6.719	7.524	7.910	46.692	47.539	53.868
68	Er	1.405	1.443	(1.643)	6.947	7.809	8.188	48.213	49.119	55.672
69	Tm	1.462	1.503		7.179	8.100	8.467	49.764	50.733	57.506
70	Yb	1.521	1.567	(1.765)	7.414	8.400	8.757	51.345	52.380	59.356
71	Lu	1.581	1.631	(1.832)	7.654	8.708	9.047	52.956	54.061	61.272
72	Hf	1.644	1.697	(1.894)	7.898	9.021	9.346	54.602	55.781	63.222
73	Ta	1.709	1.765	(1.964)	8.145	9.342	9.650	56.267	57.523	65.212
74	W	1.774	1.835	(2.035)	8.396	9.671	9.960	57.972	59.308	67.233
75	Re	1.842	1.906	(2.106)	8.651	10.008	10.274	59.708	61.130	69.298
76	Os	1.914	1.978	(2.182)	8.910	10.354	10.597	61.476	62.990	71.401
77	Ir	1.978	2.053	(2.254)	9.174	10.706	10.919	63.276	64.885	73.548
78	Pt	2.048	2.127	(2.331)	9.441	11.069	11.249	65.112	66.821	75.735
79	Au	2.121	2.204	(2.409)	9.712	11.440	11.583	66.978	68.792	77.971
80	Hg	2.195	2.282	(2.487)	9.987	11.821	11.922	68.883	70.807	80.240

202 *Chemical Microanalysis Using Electron Beams*

Table A.3   Characteristic X-ray energies (keV). In general the more prominent lines only are listed. Numbers in parentheses refer to weak lines listed in order to complete a particular column. (Taken from Johnson and White by kind permission of ASTM.)

Notes: 1. $K_{\alpha 2} = 1/3 \, K_{\alpha} + 2/3 \, K_{\alpha 1}$ (relative weighting) - e.g. Ni $K_{\alpha 2} = 1/3 \times 7.460 + 2/3 \times 7.477 = 7.471$ keV as used elsewhere in this book (see e.g. Example calculation 4.10).

For $Z = 10\text{-}19$ (Ne-K) $K_{\alpha} = K_{\alpha 2}$.

For $Z = 18\text{-}30$ (Ar-Zn) $K_{\beta 1}$ includes $K_{\beta 3}$.

2. For $Z = 40\text{-}70$ (Zr-Yb) $L_{\beta 2}$ includes $L_{\beta 15}$.

For $Z = 20\text{-}36$ (Ca-Kr) $L\alpha_1$ includes $L\alpha_2$.

2. From $Z = 83$ onwards $L_{\beta 1}$ and $L_{\beta 2}$ interchange energy-wise.

3. For $Z = 77\text{-}92$ (Ir-U) $M_{\alpha} = M_{\alpha 1} + M_{\alpha 2}$. The energy is a straight average.

Z	El	M			L			K			El	Z
		M_α	M_β	M_γ	L_α	$L_{\beta 1}$	$L_{\beta 2}$	$K_{\alpha 2}$	K_α	$K_{\beta 1}$		
81	Tl	2.268	2.362	(2.570)	10.257	12.211	12.270	70.820	72.859	82.562	Tl	81
82	Pb	2.342	2.442	(2.652)	10.550	12.612	12.621	72.792	74.956	84.922	Pb	82
83	Bi	2.419	2.525		10.837	12.978	13.021	74.802	77.095	87.328	Bi	83
84	Po	2.505	2.601	(2.735)	11.129	13.445	13.338	76.851	79.279	89.781	Po	84
85	At	2.584	2.682		11.425	13.874		78.930	81.499	92.287	At	85
86	Rn	2.665	2.766		11.725	14.313		81.051	83.768	94.850	Rn	86
87	Fr	2.747	2.852		12.029	14.768	14.448	83.217	86.089	97.460	Fr	87
88	Ra	2.830	2.940		12.338	15.233	14.839	85.419	88.454	100.113	Ra	88
89	Ac	2.914	3.033		12.650	15.710		87.660	90.868	102.829	Ac	89
90	Th	2.991	3.145	(3.369)	12.967	16.199	15.621	89.938	93.334	105.591	Th	90
91	Pa	3.077	3.239	(3.465)	13.288	16.699	16.022	92.271	95.852	108.409	Pa	91
92	U	3.165	3.336	(3.563)	13.612	17.217	16.425	94.649	98.422	111.281	U	92

A.4 Fluorescence Yields

Z	Element	ω_K	ω_{L1}^*	ω_{L2}^*	ω_{L3}
5	B	1.7×10^{-3}			
6	C	2.8×10^{-3}			
7	N	5.2×10^{-3}			
8	O	8.3×10^{-3}			
9	F	0.013			
10	Ne	0.018			
11	Na	0.023			
12	Mg	0.030	1.2×10^{-3}	1.2×10^{-3}	1.2×10^{-3}
13	Al	0.039	7.5×10^{-4}	7.5×10^{-4}	7.5×10^{-4}
14	Si	0.050	3.9×10^{-4}	3.7×10^{-4}	3.8×10^{-4}
15	P	0.063	3.3×10^{-4}	3.1×10^{-4}	3.1×10^{-4}
16	S	0.078	3.2×10^{-4}	2.6×10^{-4}	2.6×10^{-4}
17	Cl	0.097	3.5×10^{-4}	2.4×10^{-4}	2.4×10^{-4}
18	Ar	0.118	3.8×10^{-4}	2.2×10^{-4}	2.2×10^{-4}
19	K	0.140	4.9×10^{-4}	2.7×10^{-4}	2.7×10^{-4}
20	Ca	0.163	6.1×10^{-4}	3.3×10^{-4}	3.3×10^{-4}
21	Sc	0.188	1.2×10^{-3}	8.4×10^{-4}	8.4×10^{-4}
22	Ti	0.214	1.8×10^{-3}	1.5×10^{-3}	1.5×10^{-3}
23	V	0.243	2.9×10^{-3}	2.6×10^{-3}	2.6×10^{-3}
24	Cr	0.275	4.0×10^{-3}	3.7×10^{-3}	3.7×10^{-3}
25	Mn	0.308	5.2×10^{-3}	5.0×10^{-3}	5.0×10^{-3}
26	Fe	0.340	6.5×10^{-3}	6.3×10^{-3}	6.3×10^{-3}
27	Co	0.373	7.8×10^{-3}	7.7×10^{-3}	7.7×10^{-3}
28	Ni	0.406	9.2×10^{-3}	8.9×10^{-3}	9.3×10^{-3}
29	Cu	0.440	0.011	0.010	0.011
30	Zn	0.474	0.012	0.011	0.012
31	Ga	0.507	0.013	0.012	0.013
32	Ge	0.535	0.014	0.014	0.015
33	As	0.562	0.015	0.015	0.016
34	Se	0.589	0.017	0.017	0.018
35	Br	0.618	0.020	0.020	0.020
36	Kr	0.643	0.022	0.022	0.022
37	Rb	0.667	0.024	0.025	0.024
38	Sr	0.690	0.026	0.027	0.026
39	Y	0.710	0.028	0.030	0.028
40	Zr	0.730	0.031	0.032	0.031
41	Nb	0.747	0.034	0.036	0.034
42	Mo	0.765	0.036	0.039	0.037

Z	Element	ω_K	ω_{L1}^*	ω_{L2}^*	ω_{L3}
43	Tc	0.780	0.040	0.043	0.040
44	Ru	0.794	0.043	0.046	0.043
45	Rh	0.808	0.046	0.050	0.046
46	Pd	0.820	0.049	0.054	0.049
47	Ag	0.831	0.053	0.059	0.052
48	Cd	0.843	0.058	0.065	0.056
49	In	0.853	0.062	0.070	0.060
50	Sn	0.862	0.067	0.075	0.064
51	Sb	0.870	0.072	0.080	0.069
52	Te	0.877	0.077	0.085	0.074
53	I	0.884	0.083	0.091	0.079
54	Xe	0.891	0.088	0.096	0.085
55	Cs	0.897	0.094	0.104	0.091
56	Ba	0.902	0.100	0.111	0.097
57	La	0.907	0.108	0.119	0.104
58	Ce	0.912	0.114	0.127	0.111
59	Pr	0.917	0.121	0.135	0.118
60	Nd	0.921	0.129	0.143	0.125
61	Pm	0.925	0.135	0.152	0.132
62	Sm	0.929	0.143	0.161	0.139
63	Eu	0.932	0.152	0.171	0.147
64	Gd	0.935	0.160	0.181	0.155
65	Tb	0.938	0.169	0.191	0.164
66	Dy	0.941	0.180	0.203	0.174
67	Ho	0.944	0.190	0.215	0.182
68	Er	0.947	0.201	0.227	0.192
69	Tm	0.949	0.210	0.239	0.201
70	Yb	0.951	0.221	0.251	0.210
71	Lu	0.953	0.232	0.264	0.220
72	Hf	0.955	0.243	0.277	0.231
73	Ta	0.957	0.258	0.291	0.243
74	W	0.958	0.271	0.304	0.255
75	Re	0.959	0.284	0.318	0.268
76	Os	0.961	0.293	0.331	0.281
77	Ir	0.962	0.305	0.345	0.294
78	Pt	0.963	0.318	0.359	0.306
79	Au	0.964	0.330	0.373	0.320
80	Hg	0.965	0.345	0.387	0.333

Z	Element	ω_K	ω_{L1}^*	ω_{L2}^*	ω_{L3}
81	Tl	0.966	0.358	0.401	0.347
82	Pb	0.967	0.372	0.415	0.360
83	Bi	0.968	0.382	0.429	0.373
84	Po	0.968	0.396	0.444	0.386
85	At	0.969	0.411	0.459	0.399
86	Rn	0.969	0.422	0.474	0.411
87	Fr	0.970	0.436	0.489	0.424
88	Ra	0.970	0.448	0.503	0.437
89	Ac	0.971	0.464	0.517	0.450
90	Th	0.971	0.476	0.529	0.463
91	Pa	0.972	0.485	0.538	0.476
92	U	0.972	0.503	0.549	0.489

*ω_{L1} and ω_{L2} are *effective* fluorescence yields and include Coster-Kronig processes.

Table A.4 Fluorescence yields for K and L shells. (Adapted from Krause.[22]) For M shell data see Bambynek et al.[21]

Atomic number range	Partition factor
11-19	$a_K = 1.052 - 4.39 \times 10^{-4} Z^2$
20-29	$a_K = 0.896 - 6.575 \times 10^{-4} Z$
30-60	$a_K = 1.0366 - 6.82 \times 10^{-3} Z + 4.815 \times 10^{-5} Z^2$
27-50	$a_L = 1.617 - 0.0398 Z + 3.766 \times 10^{-4} Z^2$
51-92	$a_L = 0.609 - 1.619 \times 10^{-3} Z$ $- 0.03248 \sin(0.161(Z-51))$
	(The angle within the sine is in degrees.)
60-92	$a_M = 0.65$

where

$$a_K = \frac{I_{K_\alpha}}{I_K} \qquad I_{K_\alpha} = \text{intensity of } K_\alpha \text{ line}$$
$$I_K = \text{total intensity of K lines}$$

$$a_L = \frac{I_{L_\alpha}}{I_L}$$

$$a_M = \frac{I_{M_\alpha}}{I_M}$$

(K_α, L_α and M_α are the strongest, K, L and M series of lines, respectively.)

Table A.5 Expressions for calculating partition factors, taken from Table 1 of Schreiber and Wims.[23]

X-ray Mass Absorption Coefficients and Ionisation Energies

Z	El	C	E > E' (μ/ρ) = C K λ^nK		E' > E > K (μ/ρ) = C'K λ^n'K			K > E > L₁ (μ/ρ) = C L₁ λ^nL₁	
			K	nK	E'	C'K	n'K	L₁	nL₁
2	He	0.10727	0.0246	3.030					
3	Li	0.18894	0.0548	3.030					
4	Be	0.24604	0.1110	3.030					
5	B	0.30824	0.1880	3.030					
6	C	0.38531	0.2838	3.094					
7	N	0.45355	0.4016	3.066					
8	O	0.53268	0.5320	3.041					
9	F	0.61058	0.6854	3.019					
10	Ne	0.68419	0.8669	3.000	3.55	0.820	2.7345		
11	Na	0.75844	1.0721	2.983	4.40	1.049	2.7345	0.0633	2.835
12	Mg	0.83105	1.3050	2.967	6.40	1.314	2.7345	0.0894	2.820
13	Al	0.85946	1.5596	2.953	6.20	1.616	2.7345	0.1177	2.805
14	Si	0.91309	1.8389	2.940	5.90	1.957	2.7345	0.1487	2.790
15	P	0.96522	2.1455	2.927	6.70	2.339	2.7345	0.1893	2.775
16	S	1.01931	2.4720	2.916	7.50	2.763	2.7345	0.2292	2.760
17	Cl	1.07343	2.8224	2.905	8.40	3.232	2.7345	0.2702	2.745
18	Ar	1.12540	3.2029	2.895	9.50	3.747	2.7345	0.3200	2.730
19	K	1.17770	3.6074	2.886	11.09	4.308	2.7345	0.3771	2.730

Table A.6 Parameters for calculation of mass absorption coefficients (see p. 216 for explanation).

Z	El	C	$\left(\frac{\mu}{\rho}\right) = C K \lambda^{n_K}$ E > E'		$\left(\frac{\mu}{\rho}\right) = C'_K \lambda^{n'_K}$ E' > E > K		$\left(\frac{\mu}{\rho}\right) = C L_1 \lambda^{n_{L_1}}$ K > E > L_1		$\left(\frac{\mu}{\rho}\right) = C L_2 \lambda^{n_{L_2}}$ L_1 > E > L_2		$\left(\frac{\mu}{\rho}\right) = C L_3 \lambda^{n_{L_3}}$ L_2 > E > L_3	
			K	n_K	C'_K	n'_K	L_1	n_{L_1}	L_2	n_{L_2}	L_3	n_{L_3}
20	Ca	1.22904	4.0381	2.850	13.30	4.919	2.7345	0.4378	2.730				
21	Sc	1.25125	4.4928	2.850	13.20	5.580	2.7345	0.5004	2.730				
22	Ti	1.27473	4.9664	2.850	13.00	6.293	2.7345	0.5637	2.730				
23	V	1.29768	5.4651	2.850	12.95	7.059	2.7345	0.6282	2.730				
24	Cr	1.32009	5.9892	2.850	12.90	7.879	2.7345	0.6946	2.730				
25	Mn	1.34196	6.5390	2.850	12.60	8.756	2.7345	0.7690	2.730				
26	Fe	1.36384	7.1120	2.850	12.50	9.689	2.7345	0.8461	2.730				
27	Co	1.38555	7.7089	2.850	12.40	10.682	2.7345	0.9256	2.730				
28	Ni	1.40657	8.3328	2.850	12.40	11.734	2.7345	1.0081	2.730				
29	Cu	1.42775	8.9789	2.850	12.10	12.848	2.7345	1.0961	2.730	0.8719	2.61439		
30	Zn	1.44732	9.6586	2.850	12.00	14.025	2.7345	1.1936	2.730	0.9510	2.61439		
31	Ga	1.46620	10.3671	2.850	12.00	15.265	2.7345	1.2977	2.730	1.0428	2.61439	1.0197	2.3554
32	Ge	1.48464	11.1031	2.850	12.00	16.570	2.7345	1.4143	2.730	1.1423	2.61439	1.1154	2.3554
33	As	1.50268	11.8667	2.850	12.00	17.941	2.7345	1.5265	2.730	1.2478	2.61439	1.2167	2.3554
34	Se	1.52038	12.6578	2.850				1.6539	2.730	1.3586	2.61439	1.3231	2.3554
35	Br	1.53807	13.4737	2.850				1.7820	2.730	1.4762	2.61439	1.4358	2.3554
36	Kr	1.55452	14.3256	2.850				1.9210	2.730	1.5960	2.61439	1.5499	2.3554
37	Rb	1.57132	15.1997	2.850				2.0651	2.730	1.7272	2.61439	1.6749	2.3554
										1.8639	2.61439	1.8044	2.3554

X-ray Mass Absorption Coefficients and Ionisation Energies

			E > E' $(\mu/\rho) = C K \lambda^{n_K}$		K > E > L$_1$ $(\mu/\rho) = C L_1 \lambda^{n_{L_1}}$		L$_1$ > E > L$_2$ $(\mu/\rho) = C L_2 \lambda^{n_{L_2}}$		L$_2$ > E > L$_3$ $(\mu/\rho) = C L_3 \lambda^{n_{L_3}}$	
Z	El	C	K	n_K	L$_1$	n_{L_1}	L$_2$	n_{L_2}	L$_3$	n_{L_3}
38	Sr	1.58756	16.1046	2.850	2.2163	2.730	2.0068	2.61439	1.9396	2.3554
39	Y	1.60348	17.0384	2.850	2.3725	2.730	2.1555	2.61439	2.0800	2.3554
40	Zr	1.61942	17.9976	2.850	2.5316	2.730	2.3067	2.61439	2.2223	2.3554
41	Nb	1.63507	18.9856	2.850	2.6977	2.730	2.4647	2.61439	2.3705	2.3554
42	Mo	1.65071	19.9995	2.850	2.8655	2.730	2.6251	2.61439	2.5202	2.3554
43	Tc	1.66595	21.0440	2.850	3.0425	2.730	2.7932	2.61439	2.6769	2.3554
44	Ru	1.68097	22.1172	2.850	3.2240	2.730	2.9669	2.61439	2.8379	2.3554
45	Rh	1.69573	23.2199	2.850	3.4119	2.730	3.1461	2.61439	3.0038	2.3554
46	Pd	1.71037	24.3503	2.850	3.6043	2.722	3.3303	2.61439	3.1733	2.3554
47	Ag	1.72453	25.5140	2.850	3.8058	2.714	3.5237	2.61439	3.3511	2.3554
48	Cd	1.73824	26.7112	2.850	4.0180	2.706	3.7270	2.61439	3.5375	2.3554
49	In	1.75165	27.9399	2.850	4.2375	2.698	3.9380	2.61439	3.7301	2.3554
50	Sn	1.76481	29.2001	2.850	4.4647	2.690	4.1561	2.61439	3.9288	2.3554
51	Sb	1.77775	30.4912	2.850	4.6983	2.682	4.3804	2.61439	4.1322	2.3554
52	Te	1.79048	31.8138	2.850	4.9392	2.674	4.6120	2.61439	4.3414	2.3554
53	I	1.80291	33.1694	2.850	5.1881	2.666	4.8521	2.61439	4.5571	2.3554
54	Xe	1.81326	34.5614	2.850	5.4528	2.658	5.1037	2.61439	4.7822	2.3554
55	Cs	1.82062	35.9846	2.850	5.7143	2.650	5.3594	2.61439	5.0119	2.3554

210 *Chemical Microanalysis Using Electron Beams*

Z	El	$E > E'$ $\left(\frac{\mu}{\rho}\right) = C K \lambda^{n_K}$			$K > E > L_1$ $\left(\frac{\mu}{\rho}\right) = C L_1 \lambda^{n_{L_1}}$			$L_1 > E > L_2$ $\left(\frac{\mu}{\rho}\right) = C L_2 \lambda^{n_{L_2}}$			$L_2 > E > L_3$ $\left(\frac{\mu}{\rho}\right) = C L_3 \lambda^{n_{L_3}}$		
		C	K	n_K	C	L_1	n_{L_1}	C	L_2	n_{L_2}	C	L_3	n_{L_3}
56	Ba	1.82781	37.4406	2.850	5.9888	2.650	5.6236	2.61439	5.2470	2.3554
57	La	1.83506	38.9246	2.850	6.2663	2.650	5.8906	2.61439	5.4827	2.3554
58	Ce	1.84209	40.4430		6.5488	2.650	6.1642	2.61439	5.7234	2.3554
59	Pr	1.84913	41.9906		6.8348	2.650	6.4404	2.61439	5.9643	2.3554
60	Nd	1.85613	43.5689		7.1260	2.650	6.7215	2.61439	6.2079	2.3554
61	Pm	1.86282	45.1840		7.4279	2.650	7.0128	2.61439	6.4593	2.3554
62	Sm	1.86932	46.8342		7.7368	2.650	7.3118	2.61439	6.7162	2.3554
63	Eu	1.87564	48.5190		8.0520	2.650	7.6171	2.61439	6.9769	2.3554
64	Gd	1.88179	50.2391		8.3756	2.650	7.9303	2.61439	7.2428	2.3554
65	Tb	1.88773	51.9957		8.7080	2.650	8.2516	2.61439	7.5140	2.3554
66	Dy	1.89350	53.7885		9.0458	2.650	8.5806	2.61439	7.7901	2.3554
67	Ho	1.89909	55.6177		9.3942	2.650	8.9178	2.61439	8.0711	2.3554
68	Er	1.90446	57.4855		9.7513	2.650	9.2643	2.61439	8.3579	2.3554
69	Tm	1.90969	59.3896		10.1157	2.650	9.6169	2.61439	8.6480	2.3554
70	Yb	1.91472	61.3323		10.4864	2.650	9.9782	2.61439	8.9436	2.3554
71	Lu	1.91957	63.3138		10.8704	2.650	10.3486	2.61439	9.2441	2.3554
72	Hf	1.92376	65.3508		11.2707	2.650	10.7394	2.61439	9.5607	2.3554
73	Ta	1.92812	67.4164		11.6815	2.650	11.1361	2.61439	9.8811	2.3554

		$E > E'$ $\left(\frac{\mu}{\rho}\right) = C K \lambda^{n_K}$		$K > E > L_1$ $\left(\frac{\mu}{\rho}\right) = C L_1 \lambda^{n_{L_1}}$		$L_1 > E > L_2$ $\left(\frac{\mu}{\rho}\right) = C L_2 \lambda^{n_{L_2}}$		$L_2 > E > L_3$ $\left(\frac{\mu}{\rho}\right) = C L_3 \lambda^{n_{L_3}}$	
Z	El	C	K n_K	L_1	n_{L_1}	L_2	n_{L_2}	L_3	n_{L_3}
74	W	1.93223	69.5250	12.0998	2.650	11.5440	2.61439	10.2068	2.3554
75	Re	1.93611	71.6764	12.5267	2.650	11.9587	2.61439	10.5353	2.3554
76	Os	1.93979	73.8708	12.9680	2.650	12.3850	2.61439	10.8709	2.3554
77	Ir	1.94320	76.1110	13.4185	2.650	12.8241	2.61439	11.2152	2.3554
78	Pt	1.94643	78.3948	13.8799	2.650	13.2726	2.61439	11.5637	2.3554
79	Au	1.94943	80.7249	14.3528	2.650	13.7336	2.61439	11.9187	2.3554
80	Hg	1.95219	83.1023	14.8393	2.650	14.2087	2.61439	12.2839	2.3554
81	Tl	1.95466	85.5304	15.3467	2.650	14.6979	2.61439	12.6575	2.3554
82	Pb	1.95696	88.0045	15.8608	2.650	15.2000	2.61439	13.0352	2.3554
83	Bi	1.95909	90.5259	16.3875	2.650	15.7111	2.61439	13.4186	2.3554
84	Po	1.96083	93.1050	16.9393	2.650	16.2443	2.61439	13.8138	2.3554
85	At	1.96248	95.7299	17.4930	2.650	16.7847	2.61439	14.2135	2.3554
86	Rn	1.96395	98.4040	18.0490	2.550	17.3371	2.61439	14.6194	2.3554
87	Fr	1.96510	101.1370	18.6390	2.550	17.9065	2.61439	15.0312	2.3554
88	Ra	1.96607	103.9219	19.2367	2.550	18.4843	2.61439	15.4444	2.3554
89	Ac	1.96695	106.7553	19.8400	2.550	19.0832	2.61439	15.8710	2.3554
90	Th	1.96749	109.6509	20.4721	2.550	19.6932	2.61439	16.3003	2.3554
91	Pa	1.96786	112.6014	21.1046	2.550	20.3137	2.61439	16.7331	2.3554

212 *Chemical Microanalysis Using Electron Beams*

			E > E' $\left(\frac{\mu}{\rho}\right) = C K \lambda^{n_K}$		K > E > L$_1$ $\left(\frac{\mu}{\rho}\right) = C L_1 \lambda^{n_{L_1}}$		L$_1$ > E > L$_2$ $\left(\frac{\mu}{\rho}\right) = C L_2 \lambda^{n_{L_2}}$		L$_2$ > E > L$_3$ $\left(\frac{\mu}{\rho}\right) = C L_3 \lambda^{n_{L_3}}$	
Z	El	C	K	n_K	L$_1$	n_{L_1}	L$_2$	n_{L_2}	L$_3$	n_{L_3}
92	U	1.96808	115.6061		21.7574	2.650	20.9476	2.61439	17.1663	2.3554
93	Np	1.96796	118.6780		22.4268	2.650	21.6005	2.61439	17.6100	2.3554
94	Pu	1.96751	121.8180		23.0972	2.650	22.2662	2.61439	18.0568	2.3554

This is the end of the first part of Table A.6. For the table caption and explanatory notes see p. 216.

X-ray Mass Absorption Coefficients and Ionisation Energies

$L_3 > E > M_1$

$$\left(\frac{\mu}{\rho}\right) = CM_1 \lambda^{nM_1}$$

C	El	C	M_1	nM_1
30	Zn	1.44732	0.1359	2.600
31	Ga	1.46620	0.1581	2.600
32	Ge	1.48464	0.1800	2.600
33	As	1.50268	0.2035	2.600
34	Se	1.52038	0.2315	2.600
35	Br	1.53807	0.2565	2.600
36	Kr	1.55452	0.2850	2.600
37	Rb	1.57132	0.3221	2.600
38	Sr	1.58756	0.3575	2.600
39	Y	1.60348	0.3936	2.600
40	Zr	1.61942	0.4303	2.600
41	Nb	1.63507	0.4684	2.600
42	Mo	1.65071	0.5046	2.600
43	Tc	1.66595	0.5400	2.600
44	Ru	1.68097	0.5850	2.600
45	Rh	1.69573	0.6271	2.600
46	Pd	1.71037	0.6699	2.600
47	Ag	1.72453	0.7175	2.600

			$L_3 > E > M_1$ $\left(\frac{\mu}{\rho}\right) = CM_1\lambda^{nM_1}$		$M_1 > E > M_2$ $\left(\frac{\mu}{\rho}\right) = CM_2\lambda^{nM_2}$		$M_2 > E > M_3$ $\left(\frac{\mu}{\rho}\right) = CM_3\lambda^{nM_3}$		$M_3 > E > M_4$ $\left(\frac{\mu}{\rho}\right) = CM_4\lambda^{nM_4}$		$M_4 > E > M_5$ $\left(\frac{\mu}{\rho}\right) = CM_5\lambda^{nM_5}$		$M_5 > E > N_1$ $\left(\frac{\mu}{\rho}\right) = CN_1\lambda^{nN_1}$	
Z	El	C	M_1	nM_1	M_2	nM_2	M_3	nM_3	M_4	nM_4	M_5	nM_5	N_1	nN_1
48	Cd	1.73824	0.7702	2.600										
49	In	1.75165	0.8256	2.600										
50	Sn	1.76481	0.8838	2.600										
51	Sb	1.77775	0.9437	2.600										
52	Te	1.79048	1.0060	2.600										
53	I	1.80291	1.0721	2.600	0.9305	2.4471								
54	Xe	1.81326	1.1400	2.600	0.9990	2.4471								
55	Cs	1.82062	1.2171	2.600	1.0650	2.4471	0.9976	2.4471						
56	Ba	1.82781	1.2928	2.600	1.1367	2.4471	1.0622	2.4471	0.7961	2.4				
57	La	1.83506	1.3613	2.600	1.2044	2.4471	1.1234	2.4471	0.8485	2.4				
58	Ce	1.84209	1.4346	2.600	1.2728	2.4471	1.1854	2.4471	0.9013	2.4				
59	Pr	1.84913	1.5110	2.600	1.3374	2.4471	1.2422	2.4471	0.9511	2.4				
60	Nd	1.85613	1.5753	2.600	1.4028	2.4471	1.2974	2.4471	0.9999	2.4				
61	Pm	1.86282	1.6540	2.575	1.4714	2.4471	1.3569	2.4471	1.0515	2.4	1.0269	2.2	0.3300	2.498
62	Sm	1.86932	1.7228	2.575	1.5407	2.4471	1.4198	2.4471	1.1060	2.4	1.0802	2.2	0.3457	2.492
63	Eu	1.87564	1.8000	2.575	1.6139	2.4471	1.4806	2.4471	1.1606	2.4	1.1309	2.2	0.3602	2.485
64	Gd	1.88179	1.8808	2.575	1.6883	2.4471	1.5440	2.4471	1.2172	2.4	1.1852	2.2	0.3758	2.479
65	Tb	1.88773	1.9675	2.575	1.7677	2.4471	1.6113	2.4471	1.2750	2.4	1.2412	2.2	0.3979	2.472

X-ray Mass Absorption Coefficients and Ionisation Energies

Z	El	C	$L_3 > E > M_1$ $(\mu/\rho) = CM_1\lambda^{nM_1}$		$M_1 > E > M_2$ $(\mu/\rho) = CM_2\lambda^{nM_2}$		$M_2 > E > M_3$ $(\mu/\rho) = CM_3\lambda^{nM_3}$		$M_3 > E > M_4$ $(\mu/\rho) = CM_4\lambda^{nM_4}$		$M_4 > E > M_5$ $(\mu/\rho) = CM_5\lambda^{nM_5}$		$M_5 > E > N_1$ $(\mu/\rho) = CN_1\lambda^{nN_1}$	
			M_1	nM_1	M_2	nM_2	M_3	nM_3	M_4	nM_4	M_5	nM_5	N_1	nN_1
66	Dy	1.89350	2.0468	2.575	1.8418	2.4471	1.6756	2.4471	1.3325	2.4	1.2949	2.2	0.4163	2.466
67	Ho	1.89909	2.1283	2.575	1.9228	2.4471	1.7412	2.4471	1.3915	2.4	1.3514	2.2	0.4357	2.460
68	Er	1.90446	2.2065	2.575	2.0058	2.4471	1.8118	2.4471	1.4533	2.4	1.4093	2.2	0.4491	2.454
69	Tm	1.90969	2.3068	2.575	2.0898	2.4471	1.8845	2.4471	1.5146	2.4	1.4677	2.2	0.4717	2.448
70	Yb	1.91472	2.3981	2.575	2.1730	2.4471	1.9498	2.4471	1.5763	2.4	1.5278	2.2	0.4872	2.442
71	Lu	1.91957	2.4912	2.575	2.2635	2.4471	2.0236	2.4471	1.6394	2.4	1.5885	2.2	0.5062	2.436
72	Hf	1.92376	2.6009	2.575	2.3654	2.4471	2.1076	2.4471	1.7164	2.4	1.6617	2.2	0.5381	2.430
73	Ta	1.92812	2.7080	2.575	2.4687	2.4471	2.1940	2.4471	1.7932	2.4	1.7351	2.2	0.5655	2.425
74	W	1.93223	2.8196	2.575	2.5749	2.4471	2.2810	2.4471	1.8716	2.4	1.8092	2.2	0.5950	2.419
75	Re	1.93611	2.9317	2.575	2.6816	2.4471	2.3673	2.4471	1.9489	2.4	1.8829	2.2	0.6250	2.414
76	Os	1.93979	3.0485	2.575	2.7922	2.4471	2.4572	2.4471	2.0308	2.4	1.9601	2.2	0.6543	2.408
77	Ir	1.94320	3.1737	2.575	2.9087	2.4471	2.5507	2.4471	2.1161	2.4	2.0404	2.2	0.6901	2.403
78	Pt	1.94643	3.2960	2.575	3.0265	2.4471	2.6454	2.4471	2.2019	2.4	2.1216	2.2	0.7220	2.398
79	Au	1.94943	3.4249	2.575	3.1478	2.4471	2.7430	2.4471	2.2911	2.4	2.2057	2.2	0.7588	2.393
80	Hg	1.95219	3.5616	2.575	3.2785	2.4471	2.8471	2.4471	2.3849	2.4	2.2949	2.2	0.8003	2.388
81	Tl	1.95466	3.7041	2.575	3.4157	2.4471	2.9566	2.4471	2.4851	2.4	2.3893	2.2	0.8455	2.383
82	Pb	1.95696	3.8507	2.575	3.5542	2.4471	3.0664	2.4471	2.5856	2.4	2.4840	2.2	0.8936	2.378
83	Bi	1.95909	3.9991	2.575	3.6963	2.4471	3.1769	2.4471	2.6876	2.4	2.5796	2.2	0.9382	2.373

Table A.6

Parameters for calculation of mass absorption coefficients. Generally when $E_i > E > E_j$, $(\mu/\rho) = CE_j^{n_j}(12.3981/E)^{n_j}$ m² kg⁻¹, where E_i and E are in keV and $(12.3981/E)$ is λ in Å (see main text). For example, if $M_2 < E < M_3$, $(\mu/\rho) = C M_3 \lambda^{n_{M_3}}$. Exceptionally, for the energy range $E' > E > K$ $(\mu/\rho) = C'_K \lambda^{n_K}$. The table is divided into two parts. The first part relates to X-ray energies between the L_3 and N_1 edges and the second to X-ray energies down to the L_3 edge. (Data from Thinh and Leroux.²⁴)

Z	El	$L_3>E>M_1$ $(\mu/\rho)=CM_1\lambda^{n_{M_1}}$		$M_1>E>M_2$ $(\mu/\rho)=CM_2\lambda^{n_{M_2}}$		$M_2>E>M_3$ $(\mu/\rho)=CM_3\lambda^{n_{M_3}}$		$M_3>E>M_4$ $(\mu/\rho)=CM_4\lambda^{n_{M_4}}$		$M_4>E>M_5$ $(\mu/\rho)=CM_5\lambda^{n_{M_5}}$		$M_5>E>N_1$ $(\mu/\rho)=CN_1\lambda^{n_{N_1}}$	
		C	M_1	n_{M_1}	M_2	n_{M_2}	M_3	n_{M_3}	M_4	n_{M_4}	M_5	n_{M_5}	N_1	n_{N_1}
84	Po	1.96083	4.1494	2.575	3.8541	2.4471	3.3019	2.4471	2.7980	2.4	2.6830	2.2	0.9953	2.368
85	At	1.96248	4.3170	2.575	4.0080	2.4471	3.4260	2.4471	2.9087	2.4	2.7867	2.2	1.0420	2.364
86	Rn	1.96395	4.4820	2.575	4.1590	2.4471	3.5380	2.4471	3.0215	2.4	2.8924	2.2	1.0970	2.359
87	Fr	1.96510	4.6520	2.575	4.3270	2.4471	3.6630	2.4471	3.1362	2.4	2.9999	2.2	1.1530	2.355
88	Ra	1.96607	4.8220	2.575	4.4895	2.4471	3.7918	2.4471	3.2484	2.4	3.1049	2.2	1.2084	2.350
89	Ac	1.96695	5.0020	2.575	4.6560	2.4471	3.9090	2.4471	3.3702	2.4	3.2190	2.2	1.2690	2.346
90	Th	1.96749	5.1823	2.575	4.8304	2.4471	4.0461	2.4471	3.4908	2.4	3.3320	2.2	1.3295	2.341
91	Pa	1.96786	5.3669	2.575	5.0009	2.4471	4.1738	2.4471	3.6112	2.4	3.4418	2.2	1.3871	2.337
92	U	1.96808	5.5480	2.575	5.1822	2.4471	4.3034	2.4471	3.7276	2.4	3.5517	2.2	1.4408	2.333
93	Np	1.96796	5.7232	2.575	5.3662	2.4471	4.4347	2.4471	3.8503	2.4	3.6658	2.2	1.5007	2.328
94	Pu	1.96751	5.9329	2.575	5.5412	2.4471	4.5566	2.4471	3.9726	2.4	3.7781	2.2	1.5586	2.324

A.7 Logarithmic Integral

ln x	Li(x)	ln x	Li(x)	ln x	Li(x)
0.6	0.1927	3.1	10.049	5.6	62.525
0.7	0.4877	3.2	10.790	5.7	67.558
0.8	0.7702	3.3	11.584	5.8	73.024
0.9	1.0456	3.4	12.435	5.9	78.961
1.0	1.3179	3.5	13.348	6.0	85.413
1.1	1.5902	3.6	14.329		
1.2	1.8649	3.7	15.383		
1.3	2.1442	3.8	16.518		
1.4	2.4300	3.9	17.739		
1.5	2.7241	4.0	19.054		
1.6	3.0281	4.1	20.471		
1.7	3.3437	4.2	22.000		
1.8	3.6727	4.3	23.650		
1.9	4.0165	4.4	25.432		
2.0	4.3770	4.5	27.357		
2.1	4.7560	4.6	29.437		
2.2	5.1554	4.7	31.687		
2.3	5.5772	4.8	34.121		
2.4	6.0235	4.9	36.755		
2.5	6.4966	5.0	39.608		
2.6	6.9989	5.1	42.699		
2.7	7.5331	5.2	46.048		
2.8	8.1021	5.3	49.679		
2.9	8.7088	5.4	53.679		
3.0	9.3566	5.5	57.888		

Table A.7 The logarithmic integral.

$$Li(x) = \gamma + \ln(\ln x) + \sum_{i=1}^{\infty} \frac{(\ln x)^i}{i \cdot i!}$$

(see Abramowitz and Stegun[63] equations 5.1.3 and 5.1.10). Since in this book Li always appears in pairs, one component of the pair subtracted from the other, the constant γ may here be ignored.

Appendix B

Table B.1 Symbols and Acronyms

In alphabetical order (Roman followed by Greek; upper followed by lower case).

A	Relative atomic mass (atomic weight, if you are as old as the author)
a_H	Bohr radius
a_{nl}	Partition factor
b	Beam spreading
b_{nl}	Bethe ionisation cross-section parameter
C	(1) Coulombs
	(2) Constant in mass absorption coefficient parameterisation
C_i^a	Atomic concentration of element 'i'
C_i^m	Mass concentration of element 'i'
CRT	Cathode ray tube
c	Velocity of light
c_{nl}	Bethe ionisation cross-section parameter
d	Crystal plane spacing
E	Energy of electron (*never* voltage)
E'	Energy dividing two regions for mass absorption coefficient parameterisation
E_0	Initial energy of beam electron
E_0^{rel}	'Relativistically corrected' energy of electron
E_{nl}	Ionisation energy of nl level
E_p	Plasmon energy
EDS	Energy dispersive spectroscopy
EDX	Energy dispersive X-ray analysis
EPMA	Electron probe microanalysis / microanalyser

B.1 Symbols and Acronyms

ESC( )	Escape peak fraction
e	(1) Electronic charge
	(2) Base of natural logarithms
eV	Electron volts
F	(1) Farads
	(2) Fano factor
	(3) Overall ZAF correction factor
F_s, F_b, F_a, F_{ch}, F_{co}	Individual ZAF correction factors
FEG	Field emission gun
FWHM	Full width at half maximum
f( )	Function of overpotential used in continuum fluorescence correction
f_K( )	Function of Z appearing in fluorescent yield formula
f_x	Atomic scattering factor for X-rays
g	Reciprocal lattice vector
g	Magnitude of reciprocal lattice vector
h	(1) Planck's constant
	(2) Lenard's coefficient / k: absorption correction
I	Intensity
I^f	Fluoresced X-ray intensity
J	(1) Joules
	(2) Mean Bethe excitation energy
j	Total angular momentum quantum number
K	(1) First principal quantum shell (n = 1)
	(2) Electron wave vector
K⁺	High energy side of K edge
K⁻	Low energy side of K edge
k	(1) Path length exponential decay factor: absorption correction
	(2) Wave vector of electrons
L	Second principal quantum shell (n = 2)

Li	Logarithmic integral
l	Orbital quantum number
ln	Natural logarithm
M	Third principal quantum shell ($n = 3$)
MCA	Multichannel analyser
m	Magnetic quantum number
m_0	Rest mass of electron
mac	Mass absorption coefficient
N	(1) Fourth principal quantum shell ($n = 4$)
	(2) Number of K ionisations in an EDX detector
N_0	Avogadro's number
N^m	Atoms per unit mass
N^v	Atoms per unit volume
n	(1) Principal quantum number
	(2) Various integers
n_{nl}	(1) Exponent in mac parameterisation
	(2) Total number of nl ionisations: stopping power correction
P_n	Probability of creating n plasmons
PET	Pentaerythritol (WDX crystal)
PIXE	Proton induced X-ray excitation
q	Scattering vector
q	Magnitude of scattering vector
R	(1) X-ray emission efficiency factor due to electron backscattering
	(2) Radius of Rowland circle
r	Absorption edge jump ratio
S	Stopping power
SEM	Scanning electron microscope/microscopy
STEM	Scanning transmission electron microscope/microscopy
s	(1) Electron path length
	(2) Deviation from Bragg position for electron diffraction

B.1 Symbols and Acronyms

TEM	Transmission electron microscope/microscopy
t	(1) Time
	(2) Specimen thickness
	(3) Thickness of components of EDX detector
$U_{nl\ 0}$	Overpotential of beam electron with respect to level nl
U_{nl}	Initial overpotential of beam electron with respect to level nl
u	Ratio of mac's used in continuum fluorescence correction
V	Volts
v	Electron velocity
WDS	Wavelength dispersive spectroscopy
WDX	Wavelength dispersive X-ray analysis
$\overline{X}_{nl}$	Average X_{nl}
X_{nl}	E_{nl}/J
XPS	X-ray photoelectron spectroscopy
XRF	X-ray fluorescence
x	Various distances
$\overline{Z}$	Atomic number
	Average Z
ZAF	Atomic number - absorption - fluorescence X-ray intensity corrections
z	Depth below specimen surface
z_{nl}	Number of electrons in level nl
z_p	Number of valence electrons/atom
α	(1) Geometrical factor used in absorption correction for thin foils
	(2) Fine structure constant
$\alpha,\beta,\gamma...$	As subscripts, series of X-ray lines
β	v/c
ΔE	Energy difference

ε_0	Permittivity of free space
$\eta()$	Efficiency factor in EELS
θ	Scattering and various other angles
θ_c	Cut-off angle for plasmon scattering
θ_D	Detector take-off angle
θ_E	Characteristic angle for plasmon scattering
θ_s	Specimen tilt
Λ	Plasmon excitation length
λ	Wavelength
μ	X-ray absorption coefficient
$\left(\dfrac{\mu}{\rho}\right)_B^A$	Mass absorption coefficient (mac) for B X-rays by A
ν	X-ray frequency
ξ_g	Extinction distance for reflection **g**
ρ	Density
σ	(1) Scattering cross-section (various) (2) Lenard coefficient
σ_{nl}	Ionisation cross-section for level nl
τ	Pulse processing time in EDX
ϕ	Various angles
$\phi()$	Generation function: absorption correction
χ	Angle corrected mac: absorption correction
Ω	Solid angle
ω_{nl}	Fluorescent yield for level nl

I was horrified when I compiled this list to discover how many times I had used the same symbol for different purposes. I can only claim in mitigation that most of these uses were thrust on me by pre-existing convention. In fact the meaning of a symbol is usually evident from its context.

Appendix B
Table B.2 Summary of formulae

Formulae are quoted with their reference numbers as they appear in the text and with the relevant page numbers.

Bethe[5] ionisation formula

$$\sigma_{nl} = \frac{\pi e^4 z_{nl}}{(4\pi\varepsilon_0)^2 E_0 E_{nl}} \, b_{nl} \, \ln\left(\frac{c_{nl} E_0}{E_{nl}}\right) \quad (2.4) \quad (p.\,16)$$

Green and Cosslett[9] $b_K = 0.61$ $c_K = 1$

Powell[10] $b_K = 0.9$ $c_K = 0.65$ $4 < U_K < 25$
$b_{L23} = 0.6\text{-}0.9$ $c_{L23} = 0.6$ $4 < U_{L23} < 20$
$(Z \rightarrow)$

Paterson et al[8] $b_K = 0.6$ $c_K = 0.9$
$b_L = 0.6$ $c_L = 0.5$ $(L_1, L_2 \text{ or } L_3)$

Relativistic Bethe[12] ionisation formula

$$\sigma_{nl} = \frac{\pi e^4 z_{nl}}{(4\pi\varepsilon_0)^2 E_0^{rel} E_{nl}} \, b_{nl} \left[\ln\left(\frac{c_{nl} E_0^{rel}}{E_{nl}}\right) - \ln(1-\beta^2) - \beta^2\right] \quad (2.5) \quad (p.\,18)$$

Zaluzec:[13]

$\dot{b}_K = 0.9880 - 0.01883\, Z + 3.0666 \times 10^{-4}\, Z^2 - 2.154 \times 10^{-6}\, Z^3$ (2.7)
$c_K = 0.2821 + 0.0770\, Z - 3.807 \times 10^{-3}\, Z^2 + 8.262 \times 10^{-5}\, Z^3$ (p. 20) -
$- 4.784 \times 10^{-7}\, Z^4$

Electron velocity

$$v = c\sqrt{1 - \frac{1}{\left(1 + \frac{E}{511}\right)^2}}$$

(2.6a) (p. 20)

Bethe[5] stopping power

$$-\frac{dE}{ds} = \frac{\pi e^4 Z}{(4\pi\varepsilon_0)^2 E} N^v 2\ln\left(\frac{1.166 E}{J}\right)$$

(2.8) (p. 25)

$$J = 11.5 Z$$

Whiddington[16] stopping power

$$-\frac{dE}{dz} = \frac{3.9 \times 10^{-25} \rho}{E} \ \text{Jm}^{-1}$$

(2.9) (p. 26)

Lenard's law for electron transmission

$$\frac{I(z)}{I_0} = \exp(-\sigma\rho z)$$

(2.10) (p. 26)

Fluorescence ratio

$$\omega_K = \frac{f_K^4(Z)}{1 + f_K^4(Z)}$$

(2.11) (p. 34)

where $f_K(Z) = 0.015 + 0.327 Z - 0.64 \times 10^{-6} Z^3$

Mass absorption coefficients for a compound

$$\left(\frac{\mu}{\rho}\right) = \sum_i \left(\frac{\mu_i}{\rho_i}\right) c_i^m \qquad (2.14) \\ (p.\ 41)$$

Kramers'[27] bremsstrahlung

$$\frac{d\sigma_{br}}{dE} \alpha \frac{Z^2}{m_0 E E_0} \qquad (2.15) \\ (p.\ 45)$$

Elastic scattering of electrons

Rutherford[2]

$$\frac{d\sigma}{d\Omega} = \left(\frac{Z e^2}{8\pi\varepsilon_0 m_0 c^2} \frac{\sqrt{1-\beta^2}}{\beta^2} \frac{1}{\sin^2\frac{\theta}{2}}\right)^2 \qquad (2.16a) \\ (p.\ 47)$$

Mott[31] formula for light atoms

$$\frac{d\sigma}{d\Omega} =$$

$$\left(\frac{Z}{2\pi^2 \sqrt{1-\beta^2}\ a_H\ q^2}\right)^2 \left(1 - \left(\beta \sin\frac{\theta}{2}\right)^2 + \pi \beta \frac{Z}{137}\left(1 - \sin\frac{\theta}{2}\right)\sin\frac{\theta}{2}\right)$$

$$\ldots (2.17) \\ (p.\ 51)$$

Peak width of EDX detector

$$\text{FWHM}_{\text{stat}} = 2.355\sqrt{\varepsilon FE} \qquad (3.1)$$
(p. 70)

$$\text{FWHM}_{\text{total}} = \sqrt{\text{FWHM}_{\text{int}}^2 + \text{FWHM}_{\text{stat}}^2 + \text{FWHM}_{\text{elec}}^2} \qquad (3.2)$$
(p. 71)

Escape peak intensity

$$\text{ESC}(E) = 0.024 \left\{ 1 - 0.017\, E^{2.7345} \ln\left(1 + \frac{59}{E^{2.7345}}\right) \right\} \qquad (3.3)$$
(p. 74)

Thin specimens

Absorption correction

$$\frac{\mu_B}{\mu_A} \frac{\left\{ 1 - \exp\left(-\mu_A \dfrac{\cos\theta_s}{\sin(\theta_s + \theta_D)\cos\phi} t\right) \right\}}{\left\{ 1 - \exp\left(-\mu_B \dfrac{\cos\theta_s}{\sin(\theta_s + \theta_D)\cos\phi} t\right) \right\}} \qquad (4.3)$$
(p. 110)

Convergent beam foil thickness measurement[44]

$$\left(\frac{s}{n}\right)^2 = \left(\frac{1}{\xi_g}\right)^2 \left(\frac{1}{n}\right)^2 + \left(\frac{1}{t}\right)^2 \qquad (4.4)$$
(p. 111)

Fluorescence correction[46]

$$\frac{I_A^f}{I_A} = \frac{C_B^m A_A}{A_B} \times \frac{U_B \ln(c_B U_B)}{U_A \ln(c_A U_A)} \times \omega_B \times \frac{r_A - 1}{r_A} \times \left(\frac{\mu}{\rho}\right)_B^A \times \frac{\rho t}{2} \times$$

$$\left[0.923 - \ln\left\{\left(\frac{\mu}{\rho}\right)_B^{spec} \rho t\right\}\right] \quad \quad (4.5) \quad (\text{p. 123})$$

Bulk specimens

Background[47]

$$\frac{d\sigma_{br}}{dE} \propto \frac{Z}{E}\left\{k_1(E_0 - E) + k_2(E_0 - E)^2\right\} \quad \quad (4.6b) \quad (\text{p. 123})$$

Basic ZAF correction equation

$$C^{specimen} = \frac{I^{specimen}}{I^{element}} F \quad \quad (4.7) \quad (\text{p. 127})$$

where $F = F_s F_b F_a F_{ch} F_{co}$

Stopping power correction[18]

$$n_{nl}^{specimen} = C_i^a \frac{Z_{nl} b_{nl}}{2\bar{Z}}\left[U_{nl}^0 - 1 - \frac{\ln\left(\frac{\bar{X}_{nl}}{C_{nl}}\right)}{\bar{X}_{nl}}\left\{Li(\bar{X}_{nl} U_{nl}^0) - Li(\bar{X}_{nl})\right\}\right] \begin{array}{l}\text{nl ionisations}\\ \text{per incident}\\ \text{electron}\end{array}$$

$$\ldots (4.11a) \quad (\text{p. 132})$$

$$\bar{Z} = \sum_i C_i^a Z_i$$

$$\ln \bar{X} = \frac{\sum_i C_i^a Z_i \ln Z_i}{\bar{Z}}$$

$$\quad \ldots (4.12)$$
(p. 132)

$$F_s = \frac{n_{nl}^{\text{element}}}{n_{nl}^{\text{specimen}}} C_i^m$$

Backscattering correction[51]

$$R = \begin{bmatrix} 1 & Z' & Z'^2 & Z'^3 & Z'^4 & Z'^5 \end{bmatrix} \times$$

$$\begin{bmatrix} 1 & 0 & 0 & 0 & 0 & 0 \\ -0.581 & 2.162 & -5.137 & 9.213 & -8.619 & 2.962 \\ -1.609 & -8.298 & 28.79 & -47.74 & 46.54 & -17.68 \\ 5.400 & 19.18 & -75.73 & 120.05 & -110.70 & 41.792 \\ -5.725 & -21.65 & 88.13 & -136.06 & 117.75 & -42.45 \\ 2.095 & 8.947 & -36.51 & 55.69 & -46.08 & 15.85 \end{bmatrix} \begin{bmatrix} 1 \\ U_o^{-1} \\ U_o^{-2} \\ U_o^{-3} \\ U_o^{-4} \\ U_o^{-5} \end{bmatrix}$$

(4.13)
(p. 137)

where $Z' = Z/10^2$ and $F_b = R^{\text{element}}/R^{\text{specimen}}$.

$$R = \sum_i C_i^m R_i \qquad (4.14)$$
(p. 137)

Absorption correction[52, 53]

$$f_a = \frac{1}{\left(1 + \frac{\chi}{\sigma}\right)\left(1 + \frac{h}{1+h}\frac{\chi}{\sigma}\right)} \qquad (4.18)$$
(p. 141)

$$h = 1.2 \, A/Z^2 \qquad (4.19)$$
$$(p. 142)$$

$$\sigma = \frac{4.5 \times 10^4}{E_0^{1.65} - E_{nl}^{1.65}} \quad m^2 kg^{-1} \qquad (4.20)$$
$$(p. 142)$$

$$h = \sum_i C_i^m h_i \qquad (4.21)$$
$$(p. 142)$$

$$F_a = f_a^{\text{element}} / f_a^{\text{specimen}}$$

Fluorescence correction (characteristic)[18]

$$\frac{I_A^f}{I_A} = \frac{C_B^m}{2} \frac{A_A}{A_B} \omega_K^B \left(\frac{U_B^0 - 1}{U_A^0 - 1} \right)^{1.67} \frac{r_A - 1}{r_A} \left(\frac{\mu}{\rho} \right)_B^A$$

$$\times \left[\frac{\ln\left\{ 1 + \frac{\left(\frac{\mu}{\rho}\right)_A^{\text{spec}} \csc \theta}{\left(\frac{\mu}{\rho}\right)_B^{\text{spec}}} \right\}}{\left(\frac{\mu}{\rho}\right)_A^{\text{spec}} \csc \theta} + \frac{\ln\left\{ 1 + \frac{\sigma}{\left(\frac{\mu}{\rho}\right)_B^{\text{spec}}} \right\}}{\sigma} \right] \qquad \begin{array}{c}(4.25)\\(p. 151)\end{array}$$

$$F_{ch} = \frac{\left(1 + \frac{I_A^f}{I_A}\right)^{\text{element}}}{\left(1 + \frac{I_A^f}{I_A}\right)^{\text{specimen}}}$$

Fluorescence correction (continuum)[55]

$$\frac{I_A^f}{I_A} = 4.34 \times 10^{-6} \frac{r_A - 1}{r_A} A_A \bar{Z} E_K^A \frac{\left(\frac{\mu}{\rho}\right)_{AK^+}^A}{\left(\frac{\mu}{\rho}\right)_{AK^+}^{spec}} \frac{\ln(1 + uU_0)}{uU_0} \quad (4.27) \text{ (p. 156)}$$

$$F_{co} = \frac{\left(1 + \frac{I_A^f}{I_A}\right)^{element}}{\left(1 + \frac{I_A^f}{I_A}\right)^{specimen}}$$

When there are absorption edges between K_A and E_0, replace

$$E_K^A \frac{\left(\frac{\mu}{\rho}\right)_{AK^+}^A}{\left(\frac{\mu}{\rho}\right)_{AK^+}^{AB}} \frac{\ln(1 + uU_0)}{uU_0}$$

by

$$\left[\frac{\left(\frac{\mu}{\rho}\right)_{AK^+}^A}{\left(\frac{\mu}{\rho}\right)_{AK^+}^{spec}} \left\{E_K^A f^A(U_0) \frac{\ln(1 + u_A U_A^0)}{u_A U_A^0} - E_K^B f^B(U_0) \frac{\ln(1 + u_B U_B^0)}{u_B U_B^0}\right\}\right.$$

$$\left. + \frac{\left(\frac{\mu}{\rho}\right)_{BK^+}^A}{\left(\frac{\mu}{\rho}\right)_{BK^+}^{spec}} \frac{E_K^B f^B(U_0) \ln(1 + u_B U_B^0)}{u_B U_B^0}\right] \Big/ f^A(U_0) \quad (4.28) \text{ (p. 157)}$$

Beam spreading (thin specimens)[59]

$$b = 8 \times 10^{-12} \sqrt{N^v} \frac{Z}{E} t^{3/2} \quad \text{metres} \qquad (5.1) \\ (p.~174)$$

Beam spreading (bulk specimens)[18]

$$b = \frac{2 \times 10^{-7}}{\rho} \left(E_0^{1.5} - E_{nl}^{1.5} \right) \quad \text{metres} \qquad (5.2) \\ (p.~176)$$

Probability of plasmon creation[11]

$$P_n = \frac{1}{n!} \left(\frac{t}{\Lambda} \right)^n \exp\left(-\frac{t}{\Lambda} \right) \qquad (5.3) \\ (p.~181)$$

Partial cross-section correction factors[11]

$$\text{Energy } \Delta E : \eta(\Delta E) = 1 - \left(1 + \frac{\Delta E}{E_{nl}} \right)^{1-s} \qquad (5.4a) \\ (p.~184)$$

$$\text{Angle } \phi : \eta(\phi) = \frac{\ln\left(1 + \left(\frac{\phi}{\theta_E} \right)^2 \right)}{\ln\left(\frac{2}{\theta_E} \right)} \qquad (5.4b) \\ (p.~184)$$

Appendix C

A reminder on atomic structure

An atom with atomic number Z consists of a compact nucleus containing Z protons and some neutrons, and of Z *orbital* electrons which form the *electronic structure* of the atom. The electrons form a series of shells, each characterised by a *principal quantum number* n. n takes integral values 1, 2, 3, etc. The lower the value of n, the smaller the orbit around the nucleus, the lower the energy of the electrons and in terms of a particle analogy, the faster they move. Each principal quantum shell is divided into a series of sub-shells, characterised by an *orbital quantum number* l. l may take values 0, 1 ..... n-1. There are two further quantum numbers: j, the total angular momentum quantum number, which includes the spin of the electron, and m_j, a magnetic quantum number. j may take values $l \pm 1/2$, but must be positive, and m_j may take the $2j + 1$ values from -j to +j. Thus in all there are four quantum numbers, n, l, j and m_j. n has the main influence on the electron energy, followed by l, followed by j (provided l = 0). m_j has no influence unless a magnetic field is applied, whence its name.

By using the numerical laws above, we can write out a table of possible energy levels. These levels have X-ray designations K, L, M, etc. and there are also optical designations (derived from spectroscopy) which are related to n and l. By filling electrons into energy levels, the periodic table is produced. All of these aspects are shown in the table.

The details of atomic structure can be solved exactly for a hydrogen atom only. In more complicated atoms an electron is affected by the other electrons present in the orbital structure. An electron is said, for example, to be *screened* from the nucleus by those electrons closer than it to the nucleus. Some useful results for an unscreened hydrogenic atom are:

$$\text{Radius of } n^{th} \text{ shell} = \frac{n^2}{Z} a_H \qquad (C.1)$$

(where a_H = Bohr radius)

$$= \frac{4\pi\varepsilon_0 \left(\frac{h}{2\pi}\right)^2}{m_0 e^2}$$

$$(= 53 \text{ pm})$$

234 Chemical Microanalysis Using Electron Beams

with ε_o the permittivity of free space; $\hbar = h/2\pi$; h = Planck's constant; m_o, e = rest mass, charge of electron.

$$\text{Ionisation energy of } n^{th} \text{ shell} = \frac{\hbar^2}{2\,m_o a_H^2} \frac{Z^2}{n^2} \quad (C.2)$$

$$\text{Bohr orbital velocity} = \frac{\hbar}{m_o a_H} \frac{Z}{n} = \alpha\, c\, \frac{Z}{n} \quad (C.3)$$

where $\dfrac{e^2}{4\pi\varepsilon_0 \hbar c}$ is the fine structure constant $\sim \dfrac{1}{137}$ (sometimes called α)

Note that for the special case of a hydrogen atom, only n affects the energy of a shell or *level*. Screening in more complicated atoms may be taken into account in a simple way by replacing Z by Z_{eff}.

Appendix C Atomic Structure 235

n	l	j	m_j Multiplicity	X-ray Designation	Optical Designation	Element (Z) whose outermost electron occupies this level:
1	0	1/2	2	K	1s	H(1), He(2)
2	0	1/2	2	L_1	2s	Li(3), Be(4)
	1	1/2	2	L_2	2p	B(5), C(6)
		3/2	4	L_3		N(7), O(8), F(9), Ne(10)
3	0	1/2	2	M_1	3s	Na(11), Mg(12)
	1	1/2	2	M_2	3p	Al(13), Si(14)
		3/2	4	M_3		P(15), S(16), Cl(17), Ar(18)
	2	3/2	4	M_4	3d	Sc(21), Ti(22), V(23), Cr(24), Mn(25), Fe(26), Co(27), Ni(28), Cu(29), Zn(30)
		5/2	6	M_5		
4	0	1/2	2	N_1	4s	K(19), Ca(20)
	1	1/2	2	N_2	4p	Ga(31), Ge(32), As(33), Se(34), Br(35), Kr(36)
		3/2	4	N_3		
	2	3/2	4	N_4	4d	Y(39), Zr(40), Nb(41), Mo(42), Tc(43), Ru(44), Rh(45), Pd(46), Ag(47), Cd(48),
		5/2	6	N_5		
	3	5/2	6	N_6	4f	Ce(58), Pr(59), Nd(60), Pm(61), Sm(62), Eu(63), Gd(64), Tb(65), Dy(66), Ho(67), Er(68), Tm(69), Yb(70), Lu(71)
		7/2	8	N_7		

n	l	j	m_j Multiplicity	X-ray Designation	Optical Designation	Element (Z) whose outermost electron occupies this level:
5	0	1/2	2	O_1	5s	Rb(37), Sr(38)
	1	1/2	2	O_2	5p	In(49), Sn(50),
		3/2	4	O_3		Sb(51), Te(52),
						I(53), Xe(54)
	2	3/2	4	O_4	5d	La(57), Hf(72),
		5/2	6	O_5		Ta(73), W(74),
						Re(75), Os(76),
						Ir(77), Pt(78),
						Au(79), Hg(80)
	3	5/2	6	O_6	5f	Pa(91), U(92)
		7/2	8	O_7		
	4	7/2	8	O_8	5g	
		9/2	10	O_9		
6	0	1/2	2	P_1	6s	Cs(55), Ba(56)
	1	1/2	2	P_2	6p	Tl(81), Pb(82),
						Bi(83), Po(84),
		3/2	4	P_3		At(85), Rn(86)
	2	3/2	4	P_4	6d	Ac(89), Th(90)
		5/2	6	P_5		

l = 3, 4 and 5 shells are never occupied by the ground states of stable atoms

| 7 | 0 | 1/2 | 2 | Q_1 | 7s | Fr(87), Ra(88) |

l = 1-6 shells are never occupied by the ground states of stable atoms

Table C.1 Quantum numbers, X-ray designations and optical designations. The element list is a guide only. Some rearrangement of the levels can take place: see a book on atomic structure.

INDEX

(Authors are in italics. First authors only are included.)

Absorption
 - of electrons (see also Lenard), 26
 - of X-rays - see X-rays
Absorption correction - see Correction
Absorption edge, 40, 42, 97, 129, 136, 155, 157, 175, 182, 207-216
 - jump ratio, 43, 56, 73, 148, 155
Alchemi, 179
Artefacts, spectrum, 71-75
Atomic number correction - see Correction
Atomic structure, 233-236
Auger electrons, 2, 9, 10, 30, 34, 42, 66, 72, 148
 - spectroscopy, 2, 9, 37
 - yield, 33

Background, 11, 123
 - subtraction (thin specimens), 95-102
 - subtraction (thick specimens), 123-125
Backscattered electrons, 61, 95, 107
Backscattering correction - see Correction
Bambynek, W., 34, 35
Beam spreading, 11, 173, 218
 - thin specimens, 14, 121, 173-175
 - bulk specimens, 175-176
Beer's law, 40
Berger, M.J., 131
Beryllium window, 75, 76, 96, 99, 102, 106
Bethe, H.A., 15-18, 25, 26, 45, 99, 128, 170, 171, 185
Bethe
 - energy-loss equation (stopping power), 25, 27, 54, 128, 132, 171
 - ionisation cross-section equation, 16, 19, 170
 - relativistic ionisation cross-section equation, 18, 23
Bethe-Heitler theory, 45, 99
Bloch, F., 26

Bohr, N., 14, 16
Bohr orbit, 21
Born approximation, 15, 16
Bragg scattering
 - electrons, 5, 11, 12, 54-56, 111-114, 179
 - X-rays, 37, 38, 82, 84
Bragg's law, 82
Bremsstrahlung, 11, 38, 45-47, 56, 62, 93, 95-102, 119, 123, 124
 - correction - see Correction, continuum fluorescence
Bulk specimen, 4, 5, 56, 92, 100, 121-173

Cathode ray tube, 4, 57, 92, 218
cgs units, 13
Characteristic fluorescence correction - see Correction
Characteristic X-rays - see X-rays
Chen, Z., 25
Collective interaction, 8, 12, 28
Collimator, 99
Collisions
 - collective, 8, 12
 - elastic (with atom), 7, 11, 47-54, 56, 137, 142, 171, 173, 175
 - elastic (with crystal), 7, 11, 54-56
 - inelastic (with electron), 7, 12, 89, 174, 175, 180
 - inelastic (with nucleus), 7, 11, 12, 45-47
Column - see Electron microscope
Compton scattering, 37, 38
Condenser lens, 59, 62, 92, 99
Continuum - see Bremsstrahlung
Continuum fluorescence correction - see Correction
Convergent beam electron diffraction, 111, 112
Correction
 - absorption (bulk specimen), 126, 127, 139-146, 160, 163, 171, 173
 - absorption (thin film), 107-119
 - atomic number - see stopping

237

power and backscattering corrections
- backscattering, 126, 127, 136-138, 149, 160, 163, 171
- fluorescence (characteristic), 126, 127, 146-154, 160, 163, 170, 171, 173
Correction (contd)
- fluorescence (continuum), 126, 127, 154-160, 166-171, 173
- stopping power, 125-135, 149, 160, 162, 171
Correction procedures, 4
- bulk specimen, 125-172
- thin film, 102-121
COR2, 169, 170
Coster-Kronig transition, 34
Cross-section, 13
- differential, 13, 14
- ionisation, 14-25, 56, 105, 128, 183-185, 187
- partial, 183-185, 187
- X-ray production, 9, 34, 35, 105, 128
CRT - see Cathode ray tube
Crystal, detecting - see Detecting crystal

Dead layer, 75, 76, 97, 102
Dead time, 71, 72, 78, 79, 124
Deconvolution (of EDX peaks), 97, 99, 100
Detecting crystal, 86, 88, 90, 91
Detector response, 18, 75, 76, 79-82, 99, 100, 102, 105, 106, 124
Deviation parameter, 55, 111
Differential scattering cross-section, 13, 14, 46
Dirac equation, 15, 50
Discriminator circuit, 72
Doggett, J.A., 52
Duncumb, P., 137
Dyson, N.A., 32, 33, 36, 40

Edge, absorption - see Absorption edge
Edge jump ratio - see Absorption edge jump ratio
EDS - see Energy dispersive X-ray spectroscopy

EDX - see Energy dispersive X-ray spectroscopy
EELS - see Electron energy loss spectroscopy
Efficiency factor, 184, 185, 187
Egerton, R.F., 18, 20, 185, 187
Elastic, 11
Elastic scattering - see Collisions
Electromagnetic lens, 4
Electron energy loss spectrometer, 180
Electron energy loss spectroscopy, 5, 63, 66, 67, 110, 111, 180-188
Electron energy losses, 4, 25, 27, 56, 171, 175
Electron gun, 57
Electron, interactions with solids, 7
Electron microscope
- column, 4
- scanning, 3, 4, 56-61, 66, 122, 124, 175, 188
- scanning transmission, 57, 59, 66, 67, 92, 93, 124
- transmission, 3, 18, 56, 57, 62-66, 92, 93, 124, 175, 180, 188
Electron microscopy, 2
Electron probe microanalyser, 3, 4, 57, 61, 88, 122
Electron scattering - see Collisions
Electron spin, 16
Electron trajectories, 11
Electronic noise, 68, 71, 77, 78
Energy dispersive X-ray spectrometer, 66-82
Energy dispersive X-ray spectroscopy, 5, 57, 61, 62, 66-82, 93-102, 123-125, 170, 177, 178, 181, 182, 185
Energy losses - see Electron energy losses
EPMA - see Electron probe microanalyser
Escape peak, 72-75, 79, 95, 97, 99, 124
Evans, R.D., 13, 47
Exchange interactions, 16
Extinction distance, 51

Fano factor, 68, 77
Fano plot, 16, 17, 20
Fast discriminator circuit, 72
Fe^{55} source, 71

Index

Field effect transistor, 71
Field emission gun, 57, 66
Fiori, C.E., 123
Fixed beam mode, 4, 62, 92
Fluorescence of X-rays - see X-ray fluorescence
Fluorescent yield, 33, 56, 74, 105, 148, 149, 155, 203-205
Form factor, 16
Fourier transform, 97
FRAME-C, 123
Frequency filtering, 97, 124
Full width at half maximum, 68, 71, 77-79, 100
FWHM - see Full width at half maximum

Gaussian peak, 68, 100
Germanium detectors, 66, 95
Glasses as thin film standards, 104
Gold electrode, 68, 75, 76, 97, 102
Green, M., 17, 150, 170
Green and Cosslett parameters, 17, 25, 131, 132, 134, 150

Hard collisions, 14
Hartree-Slater model of the atom, 16
Heinrich, K.F.J., 40, 127, 128, 136, 139, 142, 154, 169, 170
Heisenberg uncertainty principle, 15, 46, 68
Henke, B.L., 40
Hole spectrum, 95, 100
Humphreys, C.J., 29
Hydrogenic model, 16, 185

Impact parameter, 53
Inelastic, 11
Inelastic collisions - see Collisions
Inelastic form factor, 16
Inelastic scattering - see Collisions
Inokuti, M., 16
Intermetallic compounds as thin film standards, 104
Ionisation cross-section - see Cross-section
Ionisation energy - see Absorption edge
Iteration, 110, 127, 128, 149, 169

Kelly, P.M., 111
k-factor, 106
Kramers, H.A., 45, 100, 123, 154
Krause, M.O., 34, 35

LaB_6 filament, 57, 93
Lead cerotate, 88
Lead myristate, 88
Lead stearate, 88, 177, 178
Lenard coefficient, 26, 140, 142, 146, 147, 151
Lenard's law, 26, 140, 150, 171, 175
Lenz, F., 52
Li, Y., 95
Linear WDX spectrometer, 84
Lithium drifted silicon, 66
Lithium fluoride, 88
Logarithmic integral, 120, 130, 131, 133, 217
Love, G., 170
Low energy tails, 71, 75, 96, 97, 100

Mass absorption coefficient
- electrons, 26, 140, 142, 146, 147, 151
- X-rays, 39-43, 56, 139, 146, 148, 207-216
Mass thickness, 26, 129
MCA - see Multichannel analyser
McKinley, W.A., 50, 51
Mean excitation energy, 26
Mean free path for plasmon excitation, 181, 186
Microscope, electron - see Electron microscope
Microscope spectrum, 92, 95, 107, 124, 174
Minerals as thin film standards, 104
Monochromating crystal, 5
Monte Carlo calculations, 171, 172, 175
Mosaic crystals, 84
Moseley's law, 32, 33
Mott, N.F., 50-52, 54
Multilayer detecting crystals, 177, 178
Multichannel analyser, 68, 72, 99, 100

NBS (now NIST), 123, 169, 170
Newbury, D.E., 20

Nockolds, C., 120
Nuclear elastic scattering - see Collisions
Nuclear inelastic scattering - see Collisions

Objective
 - aperture, 92, 95
Objective (contd)
 - lens, 59, 61, 62
 - pole piece, 95
Overpotential, 16, 20, 128
 - optimum, 18

Parallel detection, 83, 180
Partition factor, 34, 56, 105, 149, 206
Paterson, J.H., 17, 18
Peak-to-background, 83, 84, 124, 178
Pentaerythritol, 88, 90, 91
PET - see Pentaerythritol
Philibert, J., 140, 142
Phi-rho-z, 140, 170
Photoelectron production, 36, 38, 39, 42, 66, 139, 146
Photomultiplier, 61
Photon, 11
PIXE, 46
Plasma, 9, 28, 180
Plasmon, 9, 10, 28, 29, 181, 186
Poisson statistics, 68
Pole piece, 95
Polishing films, 106, 119, 176, 177
Powell, C.J., 17, 20, 21, 25
Proportional counter, 5, 88-90
Proton induced X-ray excitation, 46

Quantum mechanics, 15, 45, 46

Rayleigh scattering, 37, 38
Reciprocal space, 50, 113
Reed, S.J.B., 31, 69, 72, 127, 128, 136, 139, 150, 154, 157, 175
Relativistic Bethe ionisation equation, 18
Relativistic effects, 18
Relativistically corrected kinetic energy, 19, 24
Resolution, detector, 68, 71, 77, 79, 83
Response curve of EDX detector - see Detector response

Rowland circle, 84, 86
Rubidium hydrogen phthalate, 88
Russ, J.C., 48, 127, 128, 136, 139
Rutherford formulae, 12-16, 28, 47-54, 142, 173, 174

Scanning
 - coils
 - electron microscope - see Electron microscope
 - mode, 4, 62, 92
 - transmission electron microscope - see Electron microscope
Scattering factor, 50
Scattering of electrons - see Collisions
Scattering vector, 50, 113
Schrödinger equation, 15, 50
Scintillator, 61
Secondary electrons, 57, 61
SEM - see Electron microscope
Serial detection, 83, 180
Sevov, S., 170
Si(Li) detectors, 66, 89, 95, 96, 99, 100
Silicon
 - crystal, 5
 - dead layer - see Dead layer
 - diode, 61
Single electron excitation, 12
Single scattering approximation, 173
Soft collisions, 14
Soft X-rays - see X-rays
Sommerfeld, A., 45
Spatial resolution, 4, 172-176
 - effect of fluorescence on, 176
Specimen thickness - see Thin film specimens
Spectrometer
 - X-ray - see Energy dispersive and wavelength dispersive X-ray spectrometers and spectroscopies
 - electron energy loss - see Electron energy loss spectrometer and Electron energy loss spectroscopy
Spectrum artefacts, 71-75
Spin, 16, 50
Springer, G., 156, 170
Standardless analysis, 18
Standards
 - thin film, 104, 105, 107
 - bulk, 122, 123, 127

Index

STEM - see Electron microscope, scanning transmission
Stopping power, 129, 171
 - correction - see Correction
Sum peak, 71, 72
Surface films, 5

Tail - see Low energy tail
Take-off angle, 108, 139, 148
TEM - see Electron microscope
TEM, X-ray analysis in - see Thin film X-ray analysis
Thallium hydrogen phthalate, 88
Thickness, mass - see Mass thickness
Thickness of thin film specimens - see Thin film specimens
Thin film specimens
 - absorption of X-rays - see X-ray absorption
 - correction procedure, 102-121
 - fluorescence of X-rays - see X-ray fluorescence
 - standards - see Standards, thin film
 - thickness, 106, 110-116, 119, 174, 181, 186, 187
 - X-ray analysis, 92-121
Thin window detectors, 76, 77, 81, 82, 90, 178
Thinh, T.P., 39, 40, 139, 143, 152, 158
Thomson-Whiddington law, 26
Top-hat aperture, 99
Top-hat function, 97, 98
Trajectories of electrons, 11, 12, 13, 172, 175, 176
Transmission electron microscope - see Electron microscope
Tungsten filament, 57, 93
Twigg, M.E., 119

Unfolding - see Deconvolution

Wavelength dispersive X-ray spectrometer, 82-91, 93
Wavelength dispersive X-ray spectroscopy, 5, 57, 61, 62, 82-91, 124, 125, 149, 170, 177
WDS - see Wavelength dispersive X-ray spectroscopy

WDX - see Wavelength dispersive X-ray spectroscopy
Whiddington, R., 26, 123, 132, 150, 154, 171
White, H.W., 32, 33
Williams, D.J., 92, 98, 100, 106, 120
Wilson, R.R., 131
Windowless detectors, 77, 178

XPS, 37
X-ray absorption, 9, 10, 36-39, 56, 102, 107-119, 126, 127, 139-146, 177
 - edge - see Absorption edge
 - in thin films, 107-119, 122
X-ray fluorescence, 9, 10, 37, 42-44, 56, 146-160, 176
 - by characteristic X-rays, 119-120, 146-152
 - by continuum, 152-160
 - by bremsstrahlung - see X-ray fluorescence by continuum
 - in thin films, 102, 103, 107, 119, 120, 122
 - in bulk specimens, 146-160, 176
X-ray photoelectron spectroscopy, 37
X-rays
 - characteristic, 9, 30, 31, 36, 56, 68, 96, 124, 136
   (energy of) 198-202
 - cross-section for production, 9, 34, 35, 102, 105, 128
 - excitation cross-section see X-rays, cross-section for production
 - fluorescent yield - see Fluorescent yield
 - interactions with atoms, 38
 - interactions with crystals, 38
 - interactions with single core electrons, 36-38
 - interactions with single valence electrons, 37, 38
 - interactions with solids, 36-38
 - soft, 72, 141, 176-178

ZAF corrections, 122, 123, 125-172
Zaluzec, N.J., 20, 23, 25, 100

"UNTO GOD BE THE GLORY"